Glendale College Library

ENERGY WAR
Reports from the Front

ENERGY WAR
Reports from the Front

by Harvey Wasserman

☆☆☆

Lawrence Hill & Company
in association with
Meckler Publishing

333.7
E54

Library of Congress Cataloging in Publication Data

Energy war.

    Includes index.
    1. Atomic energy--Social aspects--United States--
Addresses, essays, lectures. 2. Public relations--
Atomic energy industries--United States--Addresses,
essays, lectures. I. Wasserman, Harvey.
HD9698.U52F53   333.7   79-89989
ISBN 0-88208-105-5
ISBN 0-88208-106-3 pbk.

Acknowledgements to the publications which first published many of
these articles is made where the article appears in the text.

First edition October, 1979

1 2 3 4 5 6 7 8 9 10

Lawrence Hill & Company, Publishers, Inc.
Westport, Connecticut 06880

Manufactured in the United States of America
Designed by Rainstone Designs
South Salem, NY 10590

Acknowledgements

This is dedicated to the proposition that all beings are created with certain inalienable rights, and that among them are the right to eat decent food, to drink pure water, to breathe clean air and to live in an environment free from unnatural radiation.

No one writes alone, and this is certainly no exception. Equal partners in this book have been the entire Wasserman, Shapiro and Chilewich clans; Andrea Simon; Dr. Amy S. Wainer of the Institute for Appropriate Health Care; the Montague farmers and the children of Daniel Shays; the Allen farmers and the children of Moshup; the MUSE family; Green Mountain Post; the editors and friends at the Clamshell Alliance News, *High Times*, Montana Books, *Mother Jones*, *The Nation*, *New Age*, Pacific News Service, *Politicks* (R.I.P.), the *Valley Advocate*, *WIN*, and other publications too numerous to mention; Larry Hill, Fred Jordan, Bill Whitehead, Peter Jennison, Ron Bernstein, David Bibbee, Nahum Stiskin, Hugh Van Dusen and Cindy Merman; Sara and Roy Boyd; all the good people of the alliances, concerned citizen groups, safe energy groups, ecology groups, ratepayer groups, public interest groups, consumer groups, intervenor groups, local groups and assorted other "no nukers" and servants of the sun; all the good musicians (full listing on request) who do benefit concerts and who provide the tunes when we need them; Peter Golden, Mac Holbert and all the roadies everywhere; fellow truckers J. Browne, T. Campbell and B. Hardee; the bubbas, baggers and beachers; Meldrim, Shortperson and Loeb; Kathy B., Hannah B. and Bob; Joiwind, Journey and all those of the next generation, who really do need us now; Karen Silkwood, David Comey, Michael Eakin, Paul Jacobs, Marshall Bloom and all the others who've died for our nuclear sins; our friend, the sun, and our former friend, the atom; and our collective dark side, which makes something like an energy war happen in the first place.

Contents

Preface   ix

Part One: Why the Fight Started

1   Nuclear Hazards: Visions of the Apocalypse   3
"The Nuclear Plot," *High Times*, June, 1978   3

Part Two: Reports from the Front

2   A Tower Falls (And a Movement Rises) in Montague   27
"N.O.P.E. in Mass.," *WIN* Magazine, June 27, 1974   27
"Nuke Developers on the Defensive," *WIN* Magazine,
December 3, 1974   32
"China Bomb Test Affects Valley Milk," *Valley Advocate*,
October 20, 1976   41
"Bringing the War Back Home," *Valley Advocate*, February 23, 1977   45

3   Seabrook 1976: The Shot Heard Round the World   49
"Nuclear War by the Sea," *The Nation*, September 11, 1976   49
"Trial of the Seabrook Ten," *Valley Advocate*, September 8, 1976   55
"New Battles Loom in Nuclear Controversy," *Valley Advocate*,
November 17, 1976   59
"Seabrook Nuke Funds Melted Down," *Valley Advocate*,
January 19, 1977   63
"Carter's Choice—And Ours," *New Age*, January, 1977   65

4   Seabrook 1977: The Battle Escalates   69
"High Tension in the Energy Debate—The Clamshell Response,"
The Nation, June 18, 1977   69
"The Opening Battles of the Eighties," *Mother Jones*, August, 1977   77
"The Lyndon Johnson of the Seventies?" *Valley Advocate*, June 29, 1977   85
"People Against Power," *The Progressive*, April, 1978   89
"The Power of the People: Active Nonviolence in the United States,"
New Age, September, 1977   97

5   Seabrook 1978: The Movement Hits the Mainstream   101
     "Resistance Gets Set for Spring," *The Nation*, February, 11, 1978   101
     "We Did It Again!" *Clamshell Alliance News*, July, 1978   115
     "...And Again!" *Clamshell Alliance News*, August, 1978   118
     "Power at the Polls," *Valley Advocate*, December 6, 1978   122

6   Shoot the Devil: California and Other Struggles   130
     "Nuclear Defeat in Kern County," Pacific News Service,
        March 11, 1978   131
     "The Tremors at Diablo," *The Progressive*, April, 1979   133
     "Farmers on the Line," *The Progressive*, June, 1979   138
     "An Energy Rebellion," *The Progressive*, August, 1978   143
     "Murder of Texas Nuclear Foe Stirs Suspicions,"
        Pacific News Service, April 30, 1979   147

7   Nuclear Exports: How Do You Say "Three Mile Island"
        in Tagalog?   151
     "Radioactive Roulette," *Mother Jones*, August, 1979   153

8   Japan's Nuclear Crisis—And Narita   166
     "Japan's Nuclear Crisis," *The Progressive*, November, 1976
        & "No Nukes in Japan," *WIN* Magazine, January 15, 1976   167
     "It's Not an Airport—The Struggle at Sanrizuka," *WIN* Magazine,
        November 11, 1976   174
     "Vietnam in Japan," previously unpublished, 1978   179

Part Three: On to the Sun

9   Forging Alliances: Native Americans, Farmers, Workers   189
     "The Issue of Tribal Survival," *New Age*, October, 1978   190
     "Food and Energy," *New Age*, April, 1979   195
     "Creating Jobs from Environmentalism," *Mother Jones*, June, 1978   205
     "Unionizing Ecotopia," *Mother Jones*, June, 1978   207
     "The Green Bans," *Mother Jones*, June, 1978   219

10   The Promise and Threat of Solar Power   220
     "Government's Leap into Solar: New Mexico's Power Tower Ushers
        in Solar Electric Age," Pacific News Service, October 31, 1978   240
     "New Age Farmers Launch Ark," *Politicks*, January 3, 1978   244

11   Beyond Three Mile Island   248

Many of the articles in this book have been revised, cut or amended. Introductory comments and substantial additions appear in larger type. Most of chapter 10 and all of chapter 11 appear for the first time in this book.

Preface

✩✩✩

At 3:58 A.M. on March 28, 1979, alarms in the control room at Three Mile Island Nuclear Station, south of Harrisburg, Pennsylvania, began to flash. Unit 2, online for just three months, was having another problem. The reactor had been rushed into operation at the end of 1978 to make its owner, Metropolitan Edison, eligible for a tax break. Local activists, members of the Keystone Alliance and others, warned then that the plant wasn't safe, but they were ignored.

Within weeks after Unit 2 began to generate electricity, two safety valves broke during a turbine test. On February 1, there was a leak in a throttle valve. One day later, a pump blew a seal. Four days after that, another pump tripped off. The pump was fixed, but the reason for the failure remained a mystery. Still, the utility pushed ahead.

On March 28, with Unit 2 operating at 97 percent of capacity, a generator pump shut down, threatening the reactor's cooling system. "Somebody was screwing around with some of the equipment in the feedwater system," said Edson Case of the Nuclear Regulatory Commission (NRC). "Whatever he was doing resulted in tripping the feedwater pumps off the line."

The water that normally circulates through the core to cool it and to carry heat to generate steam was confined in the reactor vessel. Heat and pressure began to rise. A valve opened and didn't close. Radioactive water gushed into a quench tank, then onto the floor of the containment structure. The emergency core cooling system went on—but an operator shut it off. A pump sent radioactive water into an auxiliary building, flooding it, and releasing lethal steam into the atmosphere. Deadly gamma radiation penetrated the walls of the containment building. Radioactive water soon found its way into the Susquehanna River. Equipment and operators failed, NRC officials later admitted, in ways never anticipated.

And then came the hydrogen bubble, threatening to blow the top off the containment building, holding central Pennsylvania hostage to all the horrors of fallout sickness, and delivering a crushing economic and psychological blow to the community at large. "It's my home," said Carl Noon, a construction worker in nearby Middletown. "Everything I've got is there. But I'll never trust it again. I'll always keep a suitcase packed."

Perhaps the rest of us should do the same. Three Mile Island took us to the brink of a catastrophe the likes of which this nation—no nation outside Japan—has ever known. It can—and probably will—happen again. Virtually the entire population of the American Northeast lives within fifty miles of an active atomic reactor. The same is true of the Chicago area and much of the rest of the United States. We all live in the shadow of a potential atomic holocaust, of which Three Mile Island was only a very small preview.

And what distinguished Three Mile Island from other atomic "events" was neither its severity nor the odds on its wiping out a major city. Windscale, England; Fermi I, Michigan; Browns Ferry, Alabama—all were comparable in size and peril.

What made Three Mile Island different was the level of public consciousness of it.

For long years before the accident a small but growing group of concerned citizens had battered away at the atomic industry. They were labeled everything from "misguided idealists" to "bloody elitists" to "kooks." But their message was clear. Alarmed by the costs and dangers of atomic power, they had spent countless hours and dollars on legal interventions, slide shows, tapes, talks, films, bake sales, leaflets, concerts—and had broken the law *en masse*—struggling to build a public awareness of the issue.

By the spring of 1979 the campaign had achieved a critical mass of its own. Reactor, weapon and waste sites such as Seabrook, New Hampshire; Rocky Flats, Colorado; Barnwell, South Carolina; Diablo Canyon, California; Trojan, Oregon; Grants, New Mexico; Delano,

Minnesota—long before TMI—had hosted major confrontations over the nuclear issue, actions clearly bound to escalate with or without a major accident.

By then, too, there was a major Hollywood film, *The China Syndrome*, which masterfully combined art with life in the nuclear age. As a portrayal of an accident at a fictional reactor, it succeeded so completely that when TMI went haywire, the film served as a "primer" on the basic issue.

Thus, with the groundwork well laid, the accident produced a quantum leap in public awareness. For the first time, we were all forced to confront the devastating reality of what a nuclear catastrophe could do to the fabric of our lives. It was not a pretty picture. The contradictions and lies from Metropolitan Edison and the Nuclear Regulatory Commission; the inability of local, state, and federal officials to cope with or even understand what was happening; the arrogant disdain for public fears displayed by energy czar James Schlesinger and other industry representatives, all began to fit a pattern in the public mind, creating rage and a willingness to act where before there had been only doubt.

On the other hand, the industry did not roll over and die. "My faith has not been shaken," said one top executive. "I had expected an accident before now. I don't think the industry is going down the drain. We are going to have to have nuclear power." With seventy reactors in operation, ninety under construction, and another two or three dozen in the advanced planning stages, the atomic power structure barely conceded anything had happened at all beyond a lot of bad publicity.

After all, the nuclear establishment embodies thirty-five years of the best hopes of the world's technological elite. It represents a proven source of electricity in the midst of a global energy crunch. It involves tens of billions of dollars in ongoing research, development, and production.

And it stands as the ultimate bastion of centralized corporate power, of the monopolization of energy by a financial and scientific priesthood beyond our control.

Fighting atomic power means taking on some of the world's most powerful corporations. But the challenge has been issued, it's gaining steam, and it's plunging this nation, and others, into a period of social confrontation far beyond anything we saw during the Vietnam War.

No other issue has ever evoked the kind of spontaneous international uproar as nuclear energy, and no other has promised as long and drawn-out a struggle—with as much at stake.

Stopping atomic power has come to mean far more than the mere elimination of a single technology. Nuclear opponents are now demanding a full-scale alternative that could revolutionize our economy —and our lives. Solar power means an energy economy based on grass-roots democracy. It could undercut not only the nuclear industry, but the very existence of Exxon and the multinational cartels that dominate the world energy market —and its political power structure. Ultimately, solar energy could be a revolutionary vehicle with which people can take charge of their own power supplies, leaving the world's richest corporations out in the cold. If there's a bottom line in this energy war, that's it.

At the same time, the prosolar movement has thus far demonstrated a unique and deep-seated commitment to peaceful action, challenging to their core the traditional modes of political battle. The adoption of disciplined, nonviolent mass action by late-seventies activists has indicated a maturity and staying power born of the work of Gandhi, Martin Luther King, and Cesar Chavez. These tactics could carry the seeds of a new social order in this country rarely dreamed possible during the raucous, polarized days of Vietnam. In its own way, nonviolent direct action poses every bit the threat to the political status quo that solar power poses to the energy cartels.

Together, they seem destined to turn this country upside down. Energy will be the issue of the 1980s and 1990s. Nothing is more basic to the nature of a society than the way it generates its power. The question now posed is, how shall we get our energy in the future? And there will be no easy answer. The middle ground is slim at best. The choice between nuclear and solar power is a choice between corporate rule and social democracy, between suicide and survival.

Reports from the Front

By the time of TMI thousands of Americans had made their choice and were acting on it. The "energy war" was already raging at nuclear sites and neighborhoods around the country and overseas.

This book is a running documentary account of part of that struggle, as it has escalated. Like the "war" itself, it is a work in progress, a collection of articles and essays that we hope will shed some light on where this confrontation has come from and where it might be headed.

We start (Chapter One) with the primary source of the struggle, the threat to health, safety, and civil liberties posed by atomic power. The proliferation of reactors threatens not only to condemn us to radioactive extinction, but to put an end also to our fundamental freedoms. If there's one basic reason why people have been willing to put their time, their money and their bodies on the line in this fight, that's it.

Putting one's life on the line for a political purpose can be a complex undertaking. In the United States, civil disobedience against nuclear energy began in the small Massachusetts town of Montague, with the toppling of a five hundred-foot weather tower by a young organic farmer named Sam Lovejoy. His defiant action against a local utility signaled a new phase in opposition to nuclear power (Chapter Two).

That opposition took on national significance one hundred twenty miles northeast of Montague, at Seabrook, New Hampshire. A series of mass arrests there brought global attention to a quiet, conservative New Hampshire town. From a handful of local opponents, the movement at Seabrook mushroomed into a confrontation of enormous historical significance to both energy and the future of social activism (Chapters Three, Four, and Five).

What was happening in New England was also happening in various forms all over the United States. In Chapter Six we look at the "war" in a few other places: in California, where a *de facto* moratorium on new construction was accompanied by a national-scale battle over the Diablo Canyon reactors, which sit near a major earthquake fault; in Minnesota, where a raucous, violent struggle is raging over high-voltage power lines; in Crystal City, Texas, where a Mexican-American community defied the energy monopolies and took charge of its own gas supply; and in Houston, where the murder of an antinuclear activist raised new fears about the real ante in this war.

In Chapter Seven we look overseas, at the issue of atomic exports to nonindustrial nations. In Chapter Eight we look at Japan, the world's second-largest generator of electricity through atomic power. Japan has more American-made reactors than any other country outside the United States. It also has an energy war of its own, one that was born not at a nuclear site but at the infamous Narita International Airport, where an unprecedented peasant struggle has been raging since the mid-sixties.

In the book's final part, "On to the Sun," we start (Chapter Nine) with a look at the alliances with Native American, farm, and labor movements that have become part and parcel of the solar push.

We then examine the problems and prospects of solar power (Chapter Ten). What are the real costs of nuclear and fossil fuels? Can we really convert to renewable energy sources? What are the social and political implications of doing it? What are the ecological and human costs of not doing it? At the end of the chapter we visit two poles of the solar dream: the centralized, high-technology "power tower" at Sandia Laboratories in New Mexico, and the small-scale, appropriately based New Alchemy Institute at Wood's Hole, Massachusetts.

Finally, in Chapter Eleven, there's a rough inventory of the state of the "war" after TMI. What can we expect over the next few years? What might the election of 1980 mean? And how can a safe energy future be won?

Like its author, this collection has rough edges. It is a mix of articles and essays, most of which were written as magazine coverage on breaking news events. We hope the mix will provide a stronger sense of immediacy than a synthesized overview might have done.

The format is also a concession to one overriding fact: this battle is being fought at a very fast pace. It's virtually impossible these days to write a word about energy without things changing between the time it goes to the printer and the time it gets published.

So at the very end we've included a list of active safe-energy organizations that can keep you posted, issue by issue, on the latest developments in your neighborhood.

We hope this effort will help in its own way to speed the arrival of a safe, sane, solarized energy.

In the meantime, we'll be seeing you on the gas lines—and on the battle lines.

No Nukes!
Harvey Wasserman
Montague, Massachusetts
Summer, 1979

ENERGY WAR
Reports from the Front

PART ONE

Why the Fight Started

"I don't know what we are protecting at this point. I think we ought to be moving people."

—Roger Mattson, Director
Nuclear Regulatory Commission
Systems Safety Division
From Three Mile Island

"Which amendment guarantees freedom of the press? I'm against it."

—Joseph Hendrie, Chairman
Nuclear Regulatory Commission

1
Nuclear Hazards:
Visions of the Apocalypse

☆☆☆

The Nuclear Plot
(*High Times*, June, 1978)

THREE MILE ISLAND introduced the American public as a whole to the dangers of a reactor catastophe.

But for the atomic industry, the accident was neither the first nor the worst.

On October 8, 1957, for example, a serious problem developed in the nuclear power plant at Windscale, on the English side of the Irish Sea. Through a complex chain of malfunctions and mistakes, some of the uranium in the reactor's fifteen hundred fuel rods caught fire.

Temperature gauges soon began flipping out of control, and radiation monitors at the top of the reactor's twin, square four hundred-foot-high emission towers sent down some bad news: radioactive gases were escaping into the countryside.

Terrified plant managers soon realized they had a nightmare on their hands. After a special scanning device malfunctioned, two technicians donned bulky radiation suits and opened an inspection hole to look in. They saw the last thing they wanted to see: more than a hundred fuel rods were glowing bright red, with blue flames surrounding them.

Summoning six more heavily suited workmen, the technicians began poking at the uranium pellets with long steel rods, trying to knock them free of the casings. It didn't work. Neither did blasts of carbon dioxide, shot in to suffocate the flames.

Throughout that day—a Thursday—steady winds blew the radioactive gases out to the Irish Sea, contaminating fish and other marine life.

That night, the wind shifted to a southerly direction. The gases began pouring inland, poisoning farmland, livestock, and people. Gamma-ray readings at Seascale, a mile away, came in abnormally high. But by now plant technicians knew there was no way to let the fire run its course. They would have to take drastic measures.

Rumors of trouble soon leaked from the plant. Panic crept through the area. One scientist fled with his family to the south of England. "There was not a large amount of radioactivity released," came the official word. "There has been no injury to any person. There is no danger of the reactor's exploding."

The disclaimer was not all that reassuring. Even as it was being made, plant technicians were debating the use of their last resort—the dumping of thousands of gallons of water into the reactor core to put out the fire.

Doing that would destroy the multimillion-dollar machine. It would also create several tons of extremely dangerous radioactive water. And, worst of all, it could cause a hydrogen explosion capable of ripping the reactor apart and shooting a murderous cloud of radioactive gases and particulates out into the atmosphere.

Meanwhile the fire burned on. Local police were put on alert for a possible attempt at a mass evacuation.

And then the decision was made. On Friday morning, October 9, a plant manager and a local fire chief dragged a fire hose up an eighty-foot ladder to the top of the reactor. At nine A.M. they turned on the water to a trickle. Technicians and firemen jumped behind steel barriers and held their breath. Live radioactive steam soared into the air. But there was no explosion. Everyone at Windscale breathed a sigh of relief. Gradually the fire hose was given full head. The reactor was dead in a day.

So were some illusions about nuclear power. Officials' reports were contradictory. Some said the stack filters had prevented all but iodine gases from escaping. Others claimed radioactive particles could also get through, including strontium, cesium, plutonium, americium, and other carcinogenic killers. The iodine alone could provoke an epidemic of thyroid cancers among children in the area; other elements could inflict a sickening array of cancers, leukemia, and birth defects that would pollute a hundred generations.

Official pronouncements continued to minimize the damage. But the government was forced to seize milk from more than a hundred farms, and soon spread the ban to a two hundred-square-mile area. Hundreds of cows, sheep, and goats were confiscated, shot, and buried. Thousands of gallons of milk were poured into the Irish Sea, where marine life got yet another dose. Farmers slaughtering their animals for meat were told to send the thyroid glands to the government for testing. Regular miners working in nearby coal shafts were replaced because radioactive doses had filtered through the ventilation systems.

And in London, three hundred miles away, the readings on radiation meters jumped significantly.

Motor City Meltdown

By this time the world was no stranger to nuclear accidents. One had already occurred six years earlier at the NRX reactor at Chalk River, Canada, where a young naval engineer named Jimmy Carter participated in the cleanup, getting his year's radiation dose in just 89 seconds.

Another Chalk River reactor, the NRU, suffered an accident of its own the same year as Windscale. Three years after that, three workmen making a fuel core adjustment at the SL-1 test reactor in Idaho were killed in a mishap that involved foul play. One worker was pinned to the ceiling of the reactor containment by an exploding fuel rod. To prevent the contamination of civilian cemetaries, parts of the three men's bodies had to be amputated and disposed of in lead caskets along with high-level wastes. Twenty-eight years later, just prior to Three Mile Island, government documents, made public for the first time, revealed that the "accident" was actually a murder-suicide. Rather than a complex scientific malfunction, the fatal event had been deliberately caused as part of a love triangle involving the staff at the plant.

Even more mysterious were the origins of a massive radioactive explosion that devastated a vast area of the Soviet Union. Generally believed to have been caused by the eruption of a high-level waste dump, the destruction covered at least seventy square miles, rendering the area permanently uninhabitable and possibly killing thousands of people. As of 1979 information about the accident remained sketchy in the West. But both the Central Intelligence Agency and emigré Russian scientists confirmed the existence of a long stretch of highway where motorists are encouraged to stay in their cars and drive through as fast as possible. The area is a "dead zone" for human life and vegetation, a science-fiction nightmare become all too real.

Equally real was a near-catastrophe that struck the Fermi fast-breeder reactor at Monroe, Michigan. Beginning on October 5, 1966, a partial meltdown at the 300-megawatt plant prompted utility officials to seriously consider the possibility of trying to evacuate Detroit, forty miles to the north. Because of the large quantities of highly volatile liquid sodium used to cool the reactor, the danger of a massive explosion was probably far greater than at any point during the Three Mile Island incident. Yet the accident remained almost totally secret.

There had, in fact, been a long and bitter public struggle over Fermi Unit I when it was first proposed. Led by Walter Reuther and Leo Goodman, the United Auto Workers (UAW) had taken Detroit Edison all the way to the U.S. Supreme Court, trying to keep the plant from being built. Despite a strong dissent from Justice William O. Douglas, construction of the plant went ahead.

But when the $150 million facility suffered its accident, hardly a word leaked to the media or the public. At the time I happened to be an editor of a daily newspaper in Ann Arbor, forty miles to the west. We had our own Associated Press wire machine, and I was also the local correspondent for the United Press International. But neither I nor anyone I know heard a word about that accident until the early seventies, when an account of it was published in John G. Fuller's *We Almost Lost Detroit*. Had Three Mile Island had its accident in 1966, we might have heard little or nothing about it either.

In 1975 yet another near-miss occurred at the Browns Ferry reactor in Alabama. Workmen inspecting a reactor wiring system lit a candle to test air flow and it set some foam insulation on fire. Before the fire was extinguished, $100 million damage had been done, and the general public had once again been taken to the brink—with only the sketchiest information about what had happened and how.

WhiteWASH

What *could* have happened was already vaguely understood by the federal government—which wasn't talking much about it. In 1956 the Atomic Energy Commission (AEC) sponsored a study on the possible damage of a "worst-case" catastrophe. Prepared by Brookhaven National Laboratory, the "WASH-740" report was finished in March, 1957, but was not released to the public. It said that an accident at a 200-megawatt reactor could kill 3,400 people outright, injure 43,000, cause $7 billion in property damage, and permanently contaminate a land mass the size of Maryland.

The report did not please the AEC, which buried it.

In July, 1957, the Engineering Research Institute of the University of Michigan did a similar study based on the Fermi plant. It concluded that 133,000 people could be exposed to a high dose of radiation from an accident there. Half those people would die. Another 181,000 people would receive a lesser dose, and many of them might also die.

That report was marked "classified" and buried.

In 1964 the AEC updated WASH-740. Again the findings were buried, and again the reasons were obvious. A worst-case accident, said the report, could kill 45,000 people, injure 100,000 and destroy $17 billion in property. Radioactive contamination could make permanently uninhabitable a land mass the size of Pennsylvania.

This report, too, was kept from the public, for nearly ten years.

Despite the gruesome statistics, the industry forged ahead, with the AEC right behind it—pushing.

Atoms for Peace

Without the atomic bomb and the government blank check written to develop it, there might never have been a nuclear industry.

During World War II, American scientists were given unlimited resources to develop a bomb. Impetus for this Manhattan Engineering District Project (as the Army named it) had come in part from Albert Einstein, who wrote a letter to Franklin Roosevelt urging such an undertaking because Einstein and other scientists believed the Nazis were doing the same thing. Einstein later deeply regretted what came to pass.

In order to build the atomic bomb, the Manhattan Project scientists first had to produce the fissionable material that went into it—

enriched uranium and the newly discovered element plutonium. To get that material, atomic reactors were refined and developed.

The experiment succeeded, as the radioactive obliteration of Hiroshima and Nagasaki amply testified.

But the production of the bombs spurred a bureaucracy of its own. After the war, in 1946, the Atomic Energy Act created the Atomic Energy Commission. By the early 1950s the AEC found itself in charge of 65,000 construction workers, a very specialized product, and an image to promote.

None of which had much to do with America's energy needs. In 1952 a special blue-ribbon panel told Harry Truman that there could be fifteen million solar-heated homes in the United States by the year 1975. With oil and gas imports a bare fraction of what they are today, there seemed ample time for serious development of the solar alternative.

But the AEC had other ideas. In 1953 Dwight Eisenhower made his famous "Atoms for Peace" speech to the United Nations. Nuclear power, he said, would be the great hope for the world's energy future. With apparently infinite radioactive resources at hand, the AEC eagerly followed suit with the line that atomic energy would make electricity "too cheap to meter."

Atoms for Peace was a malevolent combination of public relations and private economics. Coming at the height of the American development of the hydrogen bomb, it was designed to alter the image of America as the world's nuclear bully. Couched in altruistic rhetoric, the plan laid the foundations of an industry that had little indigenous market from which to build. Thus the early orders that got General Electric and Westinghouse started in the commercial reactor field came mostly from overseas, and were financed by the U.S. Export-Import Bank. This amounted to a flat federal subsidy to the new industry in the name of world peace and development.

American domestic utilities remained unenthusiastic about atomic power. Conservative by nature, the electric companies were doubtful about the ability of the reactors to generate cheap electricity, and about their safety.

In fact, doubts were so strong that the AEC virtually had to blackmail the utilities into going nuclear. It threatened that if the investor-owned electric companies didn't build them, the government would, and would then compete in the sale of electric power.

Reluctantly, the private utilities caved in. But they demanded protection from the consequences of a major accident. In 1954 they got it, in the form of the Price-Anderson Act. Designed to spur atomic development, the Act limited liability in case of a major accident to $560 million. Only $60 million of that had to be covered by the owners of the offending reactor; the rest would come out of the public treasury.

Thus less than one tenth of the potential damage from a worst-accident case as assessed by WASH-740 was insured; a mere one hundredth of it would come from the utilities building the plant. What did not come out of the taxpayers' pockets would thus be paid for by the unfortunate victims who lived downwind of the accident, and who could expect 10 percent or less on every dollar of property they lost, not to mention the inevitable health and personal costs.

Virtually every health and property insurance policy in the United States specifically excludes damage from a nuclear accident.

In 1978 John McMillan, a federal district judge sitting in North Carolina, ruled the Price-Anderson Act unconstitutional. The court, he said, "is not a bookie" and had no business estimating the odds on a nuclear accident. The U.S. Supreme Court disagreed, 9–0, and Price-Anderson remained on the books. But had Three Mile Island or any other American reactor suffered a catastrophic meltdown, its owners would remain essentially free of responsibility for those lives and property it destroyed.

The Syndrome

Throughout the insurance debate the nuclear industry adamantly maintained that its reactors cannot "blow up" like the bombs at Hiroshima and Nagasaki. Essentially, they are right.

But other types of explosions—such as from hydrogen or liquid sodium in fast breeders—are possible, and they have occurred.

And there still remains the worst-case scenarios, the China syndrome, now made a household term by the movie. The syndrome refers to a loss-of-coolant accident in which the radioactive reactor core would overheat and "melt down" into an incredibly hot, molten mass. It would then burn through the reactor pressure vessel, either blowing off the top of the containment with a steam or chemical explosion, or melting through the bottom of it, into the earth. The phrase "China syndrome" refers to the image of a core melting all the way through the planet to China.

A radioactive mass melting into the earth's crust would vent an enormous cloud of steam and lethal gases far in excess of the quantities released over Hiroshima and Nagasaki. In such a case, warns Helen Caldicott, an Australian-born pediatrician and an expert on the health effects of radiation, "thousands of people would die instantly. Thousands more would die two weeks later of acute radiation illness, where their hair falls out, their skin falls off, they get vomiting and diarrhea and they die of infection because their red and white blood cells die. Five years later there would be an epidemic of leukemia. Fifteen years later, epidemics of cancer."

In 1972 James Schlesinger, then the head of the AEC, commissioned a $3 million study into the chances of such an accident. To head it he chose Norman Rasmussen, a professor at the Massachusetts Institute of Technology and of late a director of Northeast Utilities, a major nuclear promoter.

WASH-1400, which became known as the "Rasmussen Report," took two years to complete. It said a catastrophe at a given plant could be expected once every 17,000 years. One could expect a thousand or more people to be killed in a reactor accident only once every million years.

A draft of the study became public in August of 1974 and was greeted with an almost immediate wave of scorn. Critics charged that Rasmussen's own statistics would guarantee a nation with a thousand reactors a major catastrophe every seventeen years. The Environmental Protection Agency said the report's estimate on how many people would be harmed in a meltdown was low by a factor of ten. The Union of Concerned Scientists said Rasmussen's estimate on the number of people killed outright (3300) was low by a factor of sixteen. A study by the American Physical Society indicated long-term latent cancers and genetic defects had been underestimated by a factor of fifty.

Rasmussen's study avoided assessments of the potential hazards of uranium mining, milling, and enrichment, and it did not investigate the dangers of transporting and storing radioactive wastes. It calculated the odds on a major reactor accident without factoring in the possibilities of human error. Nor did it study the potential for sabotage.

As for its methodology, physicist Amory Lovins, an energy researcher for Friends of the Earth, applied Rasmussen's model to "one particular sequence of failures in boiling water reactors." According to Lovins, "the techniques imply that the failures would occur only once in many billions of reactor years. Yet at least fifteen such accidents have already occurred in the U.S.A."

Herbert Denenberg, former insurance commissioner of Pennsylvania, concurred. "The probability of anyone believing the Rasmussen Report is one in a million," he said.

Apparently the Nuclear Regulatory Commission also came to agree. In early 1979, just prior to Three Mile Island, the Commission dropped its official endorsement of WASH-1400. It was taking no stand on the odds on a nuclear accident occurring, it said, but the Rasmussen Report no longer represented the official NRC position.

Thus TMI occurred with no official fix on the likelihood of its happening.

There were, however, some indicators. Just prior to the accident the Union of Concerned Scientists (UCS) published a book called *The Nugget File*. Using the Freedom of Information Act, the UCS obtained documents kept by Dr. Stephen H. Hanauer, a senior NRC official who reviewed "abnormal occurrences" and kept a secret file of the ones he considered most bizarre.

Among these "nuggets" was a March, 1968, incident at an unidentified plant where operators used a "regulation basketball" wrapped in tape to plug a pipe in the reactor cooling system. The pipe blew out, spilling 14,000 gallons of water into a basement within five minutes, threatening both a massive short-circuit and an uncovering of spent fuel being kept in a storage pond. According to the AEC, the problem in the pipe was solved when "a more conventional seal . . . was substituted for the basketball."

The following year testing of water samples from laboratory faucets revealed abnormal levels of radiation. Further investigation led to the discovery of a hose inexplicably connected from the drinking fountains to a 3000-gallon radioactive waste tank. Said the AEC: "The coupling of a contaminated system with a potable water system is considered poor practice in general."

That and other engineering errors prompted official AEC inspectors to remark that "in the recent past there have been a number of occurrences at reactors where human error resulted in undesirable situations. The absence of more serious effects is largely the result of good luck.''

Shake and Bake

Though luck is not something engineers generally like to bank on, there remain other ''unaccountables'' in the betting on a nuclear catastrophe.

One such imponderable comes from earthquake faults. Partly because so many reactors are located along the coasts, the nuclear industry has exhibited a strong propensity for building them on seismic shock lines. One plant—Humboldt Bay in northern California—operated for fourteen years before the NRC was forced to concede it was sited directly over a fault line, and should be shut down. Another California project—at Bodega Head—was abandoned because of seismic questions after construction had already begun, as was a double-reactor expansion planned for North Anna, Virginia. Still another California project, Diablo Canyon at San Luis Obispo, has been built two and a half miles from a major fault line that may have been active as recently as 1927 (see Chapter Six).

Just prior to TMI, the NRC ordered five East Coast reactors shut down because of the possibility of seismic damage. The action raised a howl of protest from the industry and its supporters, who charged the NRC with ''overreacting.''

But shortly after TMI an earthquake did, in fact, strike very near the Maine Yankee plant at Wiscasset, one of the five shut down. Within a month further reports surfaced from the Soviet Union indicating for the first time publicly that a reactor there had been severely damaged by seismic activity.

Nuclear construction in Iran, the Philippines, and Japan has also been sited along fault lines. It may well be that future generations will note atomic power plants as primitive markers of geologic fault lines rather than as generators of useful energy.

Low-Level Radiation

A major catastrophe is not the only reactor product that can kill large numbers of people.

Even in normal operation nuclear reactors emit low-level radiation known to cause cancer, leukemia, birth defects, premature aging, and a long list of related bodily disasters. The radiation can come in the form of rays or particles, but its effect on the human body is always deadly. "It takes one radioactive atom, one cell, and one mutated gene to kill you," warns Dr. Caldicott. "If I die of lung cancer produced by plutonium, and if I'm cremated, the smoke goes out of the chimney with the plutonium to be breathed into somebody else's lung, *ad infinitum* for half a million years."

Through the long history of atomic power, the industry has never disputed that reactors produce low-level radiation. Instead, the debate has centered on how much radioactivity is actually being produced and what quantities constitute an "acceptable" danger.

The first hints of an answer came from Dr. Alice Stewart of Oxford University. In the late fifties Dr. Stewart interviewed mothers who had been X-rayed during pregnancy. She found those women exposed in the first three months of pregnancy were twice as likely to give birth to children who suffered from leukemia before the age of ten as those mothers who were not exposed.

Statistics surfaced linking radioactive fallout to birth defects. According to Canadian government studies, birth defects rose by 78 percent in the province of Alberta for two years following Soviet bomb testing in 1958. The University of Pittsburgh's Dr. Ernest Sternglass also correlated American and Soviet atmospheric testing with a sharp increase in infant mortality after 1963.

Recent surveys of American soldiers and private citizens exposed to fallout from early bomb tests have also revealed a horrifying crop of deadly diseases, despite AEC assurances at the time that the tests were "perfectly safe."

In Pittsburgh, Dr. Sternglass carried out research covering populations living near nuclear power plants. And he came up with similar horror stories.

According to Sternglass's findings, the people living near the Shippingport, Pennsylvania, nuclear power plant have suffered abnormal rates of cancer, leukemia, and birth defects which can be traced directly to the operation of the reactor. Sternglass also contended, in a report released in the fall of 1977, that children downwind from two Northeast Utilities reactors in Connecticut received annual doses of strontium 90 more than 2.4 times greater than natural, and in quantities far greater than those recorded at other locations more distant from the plant. He further found that infant mortality rates downwind from the Millstone I reactor (at Waterford, Connecticut) rose significantly after the plant's 1970 start-up. In 1975, he concluded, citizens in three towns within thirty miles of Millstone suffered higher cancer death rates than in any other Connecticut communities.

Both Sternglass and his book, *Low Level Radiation* (published in 1971) have come under bitter industry attack. But his conclusions have held.

During the TMI crisis, for example, he warned that small children and pregnant women should be immediately evacuated from the downwind area. Industry representatives quickly branded Sternglass's warnings as those of a scaremonger.

But on the next day, Pennsylvania Governor Richard Thornburgh advised small children and pregnant women to leave the area, and later regretted he did not do more.

Sternglass's long-range conclusions have been verified by a wide array of scientists and epidemiologists, many of whom have also become targets of industry smear campaigns. Dr. Martha Drake, for example, has demonstrated abnormal cancer rates among citizens living near five relatively old nuclear reactors. Drs. Irwin Bross and Rosalie Bertell have confirmed Sternglass's links between low-level radiation and high cancer rates. Dr. Thomas Mancuso of Pittsburgh found in a study of the Hanford Nuclear Waste Storage facility in Washington that workers there suffered from abnormal cancer rates. Dr. Thomas Najarian of Boston reached similar conclusions about workers at the Portsmouth Naval Shipyard on the New Hampshire coast.

Poisoned Power

Perhaps most frightening for the general population were the findings of Drs. John Gofman and Arthur Tamplin. As two of the world's most prestigious scientists, Gofman and Tamplin were commissioned in 1963 by the AEC's Glenn Seaborg to investigate the health impact of radioactive fallout from atomic testing, and the effects of radiation from atomic reactors.

Gofman had gained world renown as a nuclear chemist and as a physician. A co-discoverer of uranium 233, he had been instrumental in the Manhattan Project that produced the Bomb, and had made scientific breakthroughs without which atomic power plants might not have been possible. Along with Arthur Tamplin, his colleague at the Lawrence Livermore Radiation Laboratory, John Gofman seemed ideally qualified to document the health effects of radiation.

In October, 1969, after six years of study, Gofman and Tamplin produced findings that indicated the AEC's radiation standards were a potential health catastrophe. If applied to every person in the United States, they would result in an additional 32,000 cancer deaths per year, and from 150,000 to 1.5 million extra genetic deaths per year in the future. As a result, Gofman and Tamplin demanded a tenfold reduction in the AEC's "permissable" dosage levels.

Despite the credentials of its authors, the study was greeted with official hostility. Like WASH-740, it posed a direct threat to the promotion of atomic power. The AEC wanted none of it. Gofman, Tamplin, and their staff were squeezed out, and their findings ignored. Says Gofman: "The AEC funded some twenty laboratories for twenty years at a rate of fifty to ninety million dollars per year to study the hazards of ionizing radiation. In that entire period there were probably fewer than ten pages of studies from the whole effort which even mentioned...possibilities of deaths from cancer from peaceful uses of the atom."

In 1971 J.I. Rodale's Organic Gardening Press of Emmaus, Pennsylvania, published Gofman and Tamplin's findings independently in a seminal volume entitled *Poisoned Power*.

Their experience with the AEC also proved seminal. Dr. Thomas Mancuso encountered official hostility to his conclusions, and even had his data confiscated. Dr. Karl Z. Morgan, founder of the health physics profession, and Drs. Irwin Bross and Rosalie Bertell all had studies confiscated or funding cut off.

None of this made nuclear power any safer, or shook the scientific conclusion that low-level radiation causes cancer, leukemia, and birth defects. All the evidence continues to indicate that there is no "safe threshold" for radiation, and that any additional dose can lead to tragedy.

In 1972 the Advisory Committee on the Biological Effects of Ionizing Radiation, a part of the National Academy of Sciences, concurred with Gofman and Tamplin that AEC exposure standards were "unnecessarily high." In April, 1979, the National Academy reconfirmed these findings, predicting "additional deaths" from the mining and milling of nuclear fuel, from the exposure of nuclear workers and the general public to reactor emissions, and from the transport, reprocessing and storage of radioactive wastes. Still more deaths, warned the report, could come from unforeseen problems such as a major accident.

Even before the "cold shutdown" of the Three Mile Island unit, Pennsylvania Governor Thornburgh charged that the NRC had "played down" the significance of radioactive doses being released from the stricken plant. This, he said, led state agencies to refrain from issuing evacuation orders, settling instead for a gubernatorial "advisory" that pregnant women and preschool children leave the immediate area.

Thornburgh was joined in his criticism by Drs. Sternglass, Morgan, and Bertell, among others.

Dr. Bertell charged that potassium iodine tablets, which block the effects of radioactive iodine, should have been distributed during the crisis. She warned that in addition to pregnant women and the very young, the elderly and mentally retarded should have been treated as citizens at high radiation risk, as well as those suffering from heart disease, arthritis, hardening of the arteries, and diabetes.

Dr. Morgan charged that the NRC never explained how it calculated those radiation doses being inflicted by the plant.

Dr. Sternglass added that those dose calculations represented just the "tip of the iceberg." Official figures comparing dosages to chest X-rays, he said, failed to acknowledge that exposure was not just to the chest but to the full body, including highly vulnerable reproductive organs. The potential damage to this and future generations, he charged, had been "wildly underestimated." Comparing the official response to the coverup of fallout dangers during the bomb testing of the fifties and sixties, Sternglass predicted a tragic epidemic of premature births, birth defects, mental retardation, leukemia, and cancer for the people of central Pennsylvania.

Only time will tell. But the outlook is not good, either for those afflicted by fallout from Three Mile Island or for those who live near other atomic reactors. Even as preliminary hearings began following the accident, deaths and stillbirths among the area's cattle were being reported, a calamity reminiscent of Windscale. "If our cows have radiation sickness, so do our wives and children," said Claire Hoover, a Middletown farmer. "It's them we worry about."

As Nobel laureate biologist George Wald put it at a Three Mile Island press conference: "Any dose of radiation is an overdose."

Mining and Wastes

Some doses can also come from the mere mining of nuclear fuel. Because it emits radioactive radon gas, uranium has brought soaring cancer rates to the industry's miners. Working in dangerous, poorly vented shafts, up to one in four uranium miners—many of them native Americans—dies of lung cancer. Lung cancer rates among uranium miners have been registered at up to ten times greater than rates even among heavy cigarette smokers living in urban areas. Some mine ventilation shafts have also been allowed to emit radon gases into native American settlements, taking still more of a human toll.

Milling the ore can also be deadly. The process leaves huge piles of radioactive "tailings," or waste material, which themselves pose a serious health threat.

In 1971 more than 90 million tons of radioactive sand were found at some thirty mills in nine western states. The government has allowed such tailings to be used in the construction of homes and schools throughout the West and South, resulting in high exposures for thousands of unsuspecting people.

The tailings also add disasterously to background radiation. A suppressed NRC study by Dr. Walter H. Jordon of the Oak Ridge National Laboratory indicates that "deaths in future generations due to cancer and genetic effects resulting from the radon from the uranium required to fuel a single reactor for one year can run into hundreds." The previous official standards for radon emissions, said Jordon, may have been lax by a factor of one hundred.

When the study became public, Jordon himself downplayed its conclusions, which led Tennessee Congressman Clifford Allen to charge "coverup." If Jordon's figures are "anywhere at all near correct," Allen told a press conference, then "one hundred reactors operating for the next twenty-five years would cause 2.5 million deaths."

The other end of the fuel cycle looks equally bleak. There is no safe, permanent depository for atomic wastes, and no immediate prospects of one. A plan to store wastes at a salt mine in Lyons, Kansas, has already failed, and no one seems to have a better idea. A high-level federal commission staged well-publicized hearings on the waste question in late 1978, but it, too, conceded that even test facilities could not be brought online until the late eighties at the earliest.

Presently, the failures are piling up. The first attempt to treat and recycle radioactive wastes began at West Valley, New York, in 1966. The plant, built some thirty miles south of Buffalo as a profit-making venture, was hailed by Governor Nelson Rockefeller as a key step toward making western New York State the "Detroit of the nuclear industry." By 1972 the plant had already changed hands, was shut by AEC order, and had dangerously polluted nearby groundwater. The taxpayers of New York were left holding 600,000 gallons of high-level residues. Minimum estimates for cleanup run to $600 million on the optimistic side, assuming someone can figure out how to do it.

Two other shots at reprocessing have also failed. General Electric poured $65 million into a facility at Morris, Illinois, which never worked right and which closed soon after it opened. A multinational consortium has also sunk $250 million into a giant reprocessing and waste-storage facility at Barnwell, South Carolina, which they hoped would be able to accept wastes from reactors all over the planet. But the builders have asked the federal government for another $750 million to finish the plant. The check is not forthcoming. There have already been numerous arrests at the facility, and since Three Mile Island the plant has encountered increasingly stiff local resistance.

Meanwhile, wastes are piling up at commercial reactors, threatening some with forced shutdowns as their capacity for storing spent fuel is used up. To avoid shutdown, plant operators have turned the reactor sites into waste dumps, with possibly disasterous consequences. Vermont Yankee at Vernon, for example, is now scheduled to place three times as many spent fuel rods in its existing storage pool as it was designed to hold. Operators intend to insert boron control rods into the pool to keep the spent fuel from fissioning and exploding. Critics charge that the storage pool could actually be more dangerous than the reactor itself. The situation is common at reactor sites around the U.S.

Even the mere transport of radioactive materials has become a major health hazard. Just prior to Three Mile Island, a truck overturned on an interstate highway in Tennessee, dumping dangerous radioactive wastes all over the road. In the 1950s the trucking of a glowing fuel rod after a Canadian reactor accident resulted in an entire highway being dug up and buried. In 1976 the wheels fell off a tractor-trailer hauling wastes south on Massachusetts Route 91. In September, 1977, a truck spilled ten thousand pounds of uranium yellowcake on a Colorado highway after it hit three horses which bolted across the road. On March 22, 1979, thousands of pounds of low-grade ore were spilled across a Kansas highway when a tractor-trailer jackknifed; some twenty-five workers were then brought in to try to clean up the highway in snow, rain, and wind. On May 4, at least ten persons were exposed to low-level radiation when a truck carrying medical wastes caught fire at a nuclear dump about 110 miles northwest of Las Vegas. By then the number of recorded accidents related to the transport of radioactive materials was well into the hundreds. At least one driver has been caught hauling radioactive materials while drunk.

It may be, however, that the French have found the ultimate answer to both the radioactive transport and the waste problem. In 1977 residents of a tiny town in France were amazed to find a large cask marked "Danger—Radiation" sitting in the middle of the town market. A call to the national authorities indicated that the cask did, indeed, contain high-level radioactive wastes.

But embarrassed officials assured the townspeople that the cask's presence in the village square did not indicate the inauguration of a new disposal facility. The cask had fallen off the truck. The driver hadn't noticed.

Sabotage and Civil Liberties

The ultimate expression of nuclear insanity is the lethal substance plutonium, miniscule quantities of which can cause cancer. A major by-product of atomic fission, plutonium was labeled "fiendishly toxic" by the AEC's Glen Seaborg. It is extremely caustic and potentially explosive. Ingesting even the tiniest microgram can be fatal.

Plutonium is also the "missing ingredient" in building a nuclear bomb, and carries with it nuclear power's ultimate threat to our individual freedoms.

In recent years the technology for making small nuclear weapons has become extremely well-known and frighteningly simple to

master. What's needed is simple engineering abilities—and plutonium, ten pounds of which leveled Nagasaki.

The U.S. General Accounting Office has estimated that at least
eight thousand pounds of plutonium currently remain unaccounted
for, enough for at least eight hundred bombs. Industry representatives
claim most of that is lodged in piping or generally diffused around
various nuclear facilities.

But at least 350 pounds of it is known to have made its way to Israel.
Sale price for a pound of plutonium on the black market is generally
estimated in the range of $10,000, meaning a full-scale bomb can be
built for roughly $100,000.

In November, 1974, Karen Silkwood, a plutonium worker and
union organizer, was killed under mysterious circumstances in Okla-
homa. She was on her way to meet with *New York Times* reporter
David Burnham, carrying documents that disappeared from her car
shortly after her death in a highway accident. Subsequent investigations strongly indicate that she may have been carrying information on
a ring of plutonium thieves operating out of the Kerr-McGee plutonium factory, where she worked.

Stolen plutonium, however, may not be the only tool for turning the
nuclear industry into a weapon of terrorism. AEC, NRC, and utility
records are filled with threats and apparent sabotage attempts at
nuclear facilities around the U.S. In the early seventies a hijacker
threatened to crash a jet into a Tennessee reactor. In May, 1979, a set
of fuel rods scheduled for use at North Anna, Virginia, were found
treated with a destructive chemical, the first publicly-confirmed sabotage of American nuclear fuel. Just prior to that, saboteurs used seven
bombs to destroy reactor components being shipped from France to
Iran. Around the same time a blue-collar worker at the La Hague
reprocessing facility in France tried to kill his boss by irradiating him
with radioactive waste, a tactic that also may have been used to
terrorize Karen Silkwood.

On October 31, 1975, the NRC published a study by John Barton of
the Stanford University law school on the civil liberties ramifications
of atomic energy. According to Barton, a known or even suspected
theft of plutonium, or the threat of sabotage at a reactor, or the threat
of malignant mass dispersal of radioactive materials in an American

city, could result in the death of Constitutional rights as we know them.

Barton's study, "Intensified Nuclear Safeguards and Civil Liberties," warned that the ante in an act or threatened act of radioactive terrorism would be so high that the Bill of Rights would not survive. "One can readily conclude," warned Barton, "that searches for the physical recovery of plutonium dispersed on the property of innocent residents will be constitutional." To find the missing materials, "authorities might attempt widespread searches of cars and pedestrians." Blanket wiretapping "would almost certainly be upheld." Known dissidents might also be "seized and detained." Their imprisonment without trial "might be justified as a way to isolate and immobilize persons capable of fashioning the material into an explosive device."

Nor would the destruction of traditional American freedoms end there. "Conceivably," wrote Barton, "detention could also be used as a step in a very troubling interrogations scheme—perhaps employing lie detectors or even torture." Indeed, concluded the report, "the chilling effect on political debate could be substantial." The fallout from nuclear terrorism could even include "torture of a person believed to know where lost material is located."

Barton's report was published by the NRC but was buried in its Washington document room until the UAW's Leo Goodman called public attention to it at a 1976 Clamshell Alliance rally. Though the report dealt with potential threats to civil liberties stemming from radioactive terrorism, antinuclear organizers have since complained of illicit wiretapping and surveillance used against them by utilities committed to nuclear power. At least three such companies have compiled extensive files on private citizens known to oppose their atomic expansion programs.

In Texas, nuclear opponents have long complained of utility surveillance, but have also been subjected to a rash of malicious attacks that may be beyond company planning. From January through May, 1979, Texas activists documented a score of beatings, tire slashings, housebreakings, car-trashings and threatening phone calls related directly to their antinuclear campaign. On April 14, two weeks after TMI, Michael Eakin, a twenty-eight year-old journalist and organizer, was shot to death on a Houston side street. His companion, Dila Davis, a worker at Balcones Research Laboratory in Austin and also an activist, took a bullet in the jaw. Though no suspects were arrested, local nuclear opponents believe the shooting was related to the atomic issue (see Chapter Six).

The Texas violence, the circumstances surrounding the death of Karen Silkwood, the utility surveillance, and heightened polarization around the nuclear issue all underscore fears that the industry and its supporters might use the extreme sensitivity of the atomic issue, or the idea that we need nuclear energy at all costs, as a tool against the opposition, and could someday use it as an excuse to infringe the civil liberties of a whole society. A nation surrounded by both atomic power plants and social instability just might find itself choosing between its basic freedoms amd radioactive chaos. "Commercial nuclear power is viable only under social conditions of absolute stability and predictability," warns Sun Day organizer Denis Hayes. "Reliance upon nuclear power as the principle source of energy is probably possible only in a totalitarian state."

Atomgate

Throughout the entire nuclear debate, the industry and its supporters have argued that our world is not risk-free, and that atomic power—like other energy sources—offers benefits that outweigh its drawbacks.

Nuclear energy, however, provides just 12 percent of our national electricity supply, just 4 percent of our total energy. It has proved not only unsafe, but expensive and unreliable as well (Chapter Ten).

More important, it carries risks to the health, safety, genetic integrity and basic freedoms of this and future generations that are a whole order of magnitude beyond anything else human society has ever confronted, short of nuclear war.

The radioactive and civil liberties nightmares made real by atomic power are simply not worth 4 percent of our nation's energy supply, especially when there are cheaper, safer, saner ways to get the power.

We don't live in a risk-free society. But we do live in one where individuals are guaranteed the right to choose which risks they'll take. And where society as a whole has the right to balance those risks against over-all benefits, with full information in hand.

As of now, a complete accounting of the health and political hazards of nuclear power cannot be made. But one basic fact is clear: the more that's learned, the worse atomic energy looks.

And the more history that leaks to the public, the more obvious it becomes that the nuclear industry has done its very best to keep the real story secret.

Don't be surprised if, at some point soon, a scandal develops over the atomic issue that entirely dwarfs what we now remember as Watergate.

In the meanwhile, as that happens, the war over nuclear energy will be raging right where it began in earnest—in the small, rural and semi-rural communities where nuclear power plants are being sited and built.

PART TWO

Reports from the Front

"If anyone had been paying attention, Three Mile Island wouldn't have happened."

—Robert Pollard
Union of Concerned Scientists

2

A Tower Falls
(And a Movement Rises)
in Montague

✩✩✩

N.O.P.E. in Mass.
(*Win* Magazine, June 27, 1974)

THE TOWN OF MONTAGUE, MASSACHUSETTS, borders the Connecticut River about ninety miles west of Boston and two hundred miles northeast of New York City. It is a classic New England collage of depressed mills, struggling small farms, and encroaching suburbs, with about 6000 of its 8555 residents living in Turners Falls, a turn-of-the-century factory center.

For as long as anyone can remember, the economy of Montague has been depressed. Unemployment is high, as are taxes. For years the town has watched neighboring Amherst to the south and Greenfield to the north expand and prosper, while the Turners Falls environs floundered in depression, apparently without hope.

Until the spring of 1973.

Then the Northeast Utilities Company (NU), which supplies much of the Connecticut Valley with its electricity, hinted it might like to build a nuclear power plant in Montague. The company said it was considering two other sites, but began wining and dining town heavies in a manner brazen enough to convince everyone they would choose Montague.

There was reason for NU to test the political waters. Townspeople had just beaten back a scheme to turn the Montague Plains, in the heart of the town, into a garbage dump. The Plains are a natural aquifer, endowed with an immense bed of sand and gravel. The Boston and Maine Railroad had wanted to excavate the gravel (worth some $600 million) and fill the hole with Boston's "sanitary landfill."

Townspeople found the plan offensive. Armed with an environmental impact study showing how the project would damage the Connecticut River, local opponents won their case.

But no sooner was the dump dumped than NU decided to drop their little A-bomb. While our selectmen—some of whom had ardently opposed the dump—visited NU's other reactors and gobbled free steak, townspeople grew enthusiastic about the influx of jobs and business and the giant boost to the tax rolls that the plant might bring.

By December, NU's announcement that Montague had been officially designated as the site for the plant was a foregone conclusion. But there was a surprise: there would be two reactors instead of one, and the cost estimate was a staggering $1.35 billion, soon raised to $1.52 billion. The reactors would go into service in 1981 and 1983. Rated at 2300 megawatts, the Montague Nuclear Power Station would be the biggest of its kind ever built.

NOPE to the "Nukes"

Despite the general enthusiasm, there was immediate opposition to the plant, divided between radical and liberal. The Montague Nuclear Concerns Group, composed mostly of university-oriented professionals, asked that the plants be built underground. When NU turned down their request, the MNCG came out of the closet and into open opposition.

Nuclear Objectors for a Pure Environment (NOPE) was more direct. A nonorganization in the 1960s tradition, NOPE immediately declared its unqualified opposition to nuclear power in general and nuclear plants in Montague in particular. The stand was taken on basic health, safety, and environmental objections. As the fight dragged on, the list grew longer.

NOPE further objected to the plant because its construction would double the population of the town, inevitably destroying the remaining farm and forest land in an area that is rapidly becoming suburbanized.

Down Goes a Tower

But that seemed precisely what was most attractive to much of Montague. The $1.52 billion represented about thirty times the entire assessed value of the town, and with the prospect of more jobs and business than the town had ever seen, vibrations were overwhelmingly favorable for the plant. NOPE was essentially the creation of a few organic farms in the rural Montague Center area, and as the town had just adopted an anticommune law, the opposition seemed isolated, to say the least.

Until George Washington's Birthday, 1974.

In the wee morning hours of February 22, Sam Lovejoy, a member of one of the local communes, slipped onto the Montague Plains and sabotaged the five hundred foot weather tower Northeast Utilities had erected to test wind direction at the site. (A company official later explained that the data was needed so that authorities would know which way the radiation would blow from the plant in case of an accident.)

Using a few simple farm tools [to loosen the turnbuckles in the stays of the tower], Lovejoy left behind him 349 feet of twisted wreckage. He then ran to the nearest road, flagged down a passing patrol car, and got a ride to the Turners Falls station, where he gave Officer Donald Cade the good news. "As a farmer concerned about the organic and the natural," he said in a typed four-page statement, "I find irradiated fruit, vegetables and meat to be inorganic; and I can find no natural balance with a nuclear plant in this or any community.

"There seems to be no way for our children to be born or raised safely in our community in the very near future. No children? No edible food? What will there be?

"While my purpose is not to provoke fear, I believe that we must act; positive action is the only option left open to us. Communities have the same rights as individuals. We must seize back control of our own community.

"The nuclear energy industry and its support elements in government are practicing actively a form of despotism. They have selected the less populated rural countryside to answer the energy needs of the cities. While not denying the urban need for electrical energy (perhaps addiction is more appropriate), why cannot reactors be built near those they are intended to serve? Is it not more efficient? Or are we witnessing a corrupt balance between population and risk?

"It is my firm conviction that if a jury of twelve impartial scientists was empanelled, and following normal legal procedure they were given all pertinent data and arguments, then this jury would never give a unanimous vote for deployment of nuclear reactors amongst the civilian population. Rather, I believe they would call for the complete shutdown of all the commercially operated nuclear plants.

"Through positive action and a sense of moral outrage, I seek to test my convictions."

A long-time resident of the town and the Valley, Sam was freed that morning on personal recognizance. Later he was indicted on one count of wanton and malicious destruction of personal property, which carried a possible five-year sentence. He pleaded "absolutely not guilty" and announced he would handle his own defense. The trial would begin in six months.

The Acid Test Begins

Response from the local community was outrage. The *Greenfield Recorder* ran a rare front-page editorial denouncing the act, and one of its columnists compared Lovejoy to John Wilkes Booth, Sirhan Sirhan, Lee Harvey Oswald, and Adolph Hitler. A local selectman wondered (in print) how Sam would like it if someone burned down his barn. Another reiterated his belief that those opposed to the plant should leave town.

But the opposition was ecstatic. Letters of support poured in, typically voicing the sentiment "I was truly inspired!" In general, the toppling of the tower forced the people of Montague and the Valley to realize that the power plant was a real issue, with more at stake than a scientific point of debate. For many it was the first realization that the plants were even planned.

In the spring, town meetings in neighboring Shutesbury, Leverett, and Wendell went on record against the plant, while the Amherst town meeting narrowly defeated a broad resolution calling for a nuclear moratorium.

But Montague was a different story. By the staggering count of 67–12 the town meeting voted down the call for a moratorium. (It also soundly defeated a resolution calling for Richard Nixon's impeachment.)

Town elections are usually held before town meeting, but this year, for the political purposes of one selectman, they came afterwards. Included on the ballot was the referendum proposition: "Should two nuclear plants be built in Montague?"

The proposition was put there by pronuclear officials who expected to win by a landslide. Indeed, nearly everyone in town predicted the issue would pass by an eight to one or ten to one ratio. One selectman predicted twenty to one.

To help focus the opposition, NOPE expanded to create the No Party, and entered a slate of five candidates—Anna Gyorgy for selectperson, Janice Frey for board of health, Marc Sills for moderator, and Lovejoy and Nina Simon for town meeting representatives from the Montague Center precinct.

Montague hadn't experienced much resembling radical politics since Shays' Rebellion, and the campaign got good press. Townspeople proved to be far less sold on the plant than had been believed. In a classic case of small-town paranoia, a family in one house would admit to being against the plant, but confess they were afraid to discuss it with their neighbors because they thought everyone was for it. Then their neighbors would admit the same thing. The campaign quickly transcended the nuke issue and offered the nuclear opponents their first real opportunity to communicate with townspeople on an individual basis.

The election results were also gratifying. The referendum passed, but the margin was less than 3 to 1. That may sound like a backhanded victory, but the fact that 770 of the 3000 voters registered solid opposition to a $1.52 billion project at such an early stage was no less than amazing. In addition, the NO candidates for town office carried more than 6 percent of the vote.

But what blew the town's mind, most of all, was that Sam Lovejoy got more than a hundred votes for town meeting member in the first precinct. Some people, it seemed, thought toppling weather towers was not such a bad thing.

☆☆☆

Nuke Developers on the Defensive
(*Win* Magazine, December 3, 1974)

SAM'S TOWER-TOPPLING got national attention, but its most important effect was on the immediate area, where "nuke" suddenly became a household word. By early spring an area-wide organization called the Alternative Energy Coalition (AEC) began a successful campaign to put a dual referendum on the state Senate District ballot, a district roughly corresponding to the Radiation Hazard Zone of the proposed plant.

The first proposition asked that the state senator be directed to oppose the Montague plant. The second asked that he be directed to "sponsor and support a resolution aimed at closing and dismantling" two active nuclear plants at Rowe, Massachusetts, and Vernon, Vermont.

By the end of the summer the AEC had gathered more than 3800 signatures, and the two questions were on the ballot. Area newspapers were beginning to fill with stories about atomic energy, and local car bumpers began to sprout a crop of blue-and-white stickers that said "No Nukes" and bore a circle with two wavy lines inside, the astrological sign of Aquarius. The symbol, said organizers, was for natural energy.

The Trial of the Tower Toppler

In early September, all eyes turned to the Franklin County Superior Court in Greenfield. Lovejoy wanted his trial to become a public forum on nuclear power, and it proved to be just that, from the opening "Hear-ye!" to the closing "Not guilty!"

At two pretrial hearings, Judge Kent Smith refused to accept that Lovejoy intended to defend himself. Confronted with a five-year felony charge, Lovejoy told the judge he considered the toppling of the tower—and the trial—a political event, and that his defending himself was inseparable from the politics of the act.

Smith, known as the most liberal judge on the Massachusetts circuit, conceded Lovejoy's right to act without counsel, but practically pleaded with him not to do it. Lovejoy held fast, but agreed to use a lawyer when it came to take the stand himself. Under Massachusetts law, Lovejoy would have had to ask himself questions, then answer them.

In pretrial motions Lovejoy asked for subpoena powers for any and all Northeast Utilities, state, or prosecution files on himself, other nuclear objectors, and on the health and safety of nuclear power plants.

Smith denied the requests, but did grant Lovejoy's "motion to view," which meant that as soon as the jury was chosen it would be bused (at Lovejoy's expense) to the site of the tower.

The actual trial began September 17th, which was both Constitution Day and the first day of Rosh Hashanah. The visitors' gallery filled with supporters, curious local citizens, and an occasional high school civics class.

But filling the courtroom proved easier than filling the jury. Nearly a quarter of the veniremen proved to have strong connections to Northeast Utilities, being either employees, relatives of employees, or stockholders. Others had already formed an opinion. By the end of the first day only twelve jurors had been chosen, and the forty-five person pool was exhausted. Judge Smith wanted two alternates, so the following morning High Sheriff Chester Martin was ordered into "the highways and byways" to collar more jurors. By 1:00 P.M. Lovejoy had used his last peremptory challenge, and the jury was completed.

After lunch, Judge Smith, defendant Lovejoy, and prosecutor John Murphy piled into Smith's huge black Buick; the fourteen jurors and the court attendants took seats on a public bus. Followed by a caravan of spectators, all went off to view the tower.

NU had flown in a replacement tower from Texas and erected it within two weeks after Lovejoy toppled the original, so all was pretty much as it had been February 22—with some notable exceptions.

For one thing, to protect the turnbuckle stations, NU had installed eight-foot storm fences topped by two-way barbed wire. Signs warned would-be topplers that the stations were surrounded by an underground alarm system. And the turnbuckles themselves were now sheathed in quarter-inch steel.

Both Lovejoy and prosecutor Murphy took pains to explain that all this hardware was not present at the time of the deed.

The sun shone brightly, the sky was a deep blue, and everyone seemed to enjoy being out on an idyllic New England fall day. The only note of discord was sounded when Lovejoy attempted to describe the fragile ecology of the Montague Plains. The prosecution objected, and was sustained.

With Malice Toward None

Back in court, prosecutor Murphy presented testimony from three NU officials and three Montague police officers, establishing beyond a doubt that the tower had been toppled, that it was worth $42,500, and that Lovejoy did it. The boredom was relieved only by the unexpected and very dramatic reading of Lovejoy's statement by Officer Donald Cade, the man on duty when Lovejoy turned himself in.

Lovejoy, however, had some fireworks ready. The charge was "willful and malicious destruction of personal property," and the core of his defense was that the act was anything but malicious, that, in fact, it was motivated by none but the highest of motives—defense of the community.

To prove his case, Lovejoy summoned Dr. John Gofman, author of *Poisoned Power*.

Lovejoy asked Gofman to tell the jury his name, address, and occupation. Lovejoy then asked Gofman to define "nuclide," whereupon Murphy stood up to object and Smith ordered the jury out of the room.

Smith asked Lovejoy to demonstrate the relevance of Gofman's testimony. Lovejoy responded that malice had not been proved by the Commonwealth and that he intended to show his motives. Gofman was the ablest person he could find to explain the dangers of nuclear power.

Smith responded that only testimony relating to Lovejoy's actual state of mind at the time of the deed would be admissible. Had he talked to Gofman before February 22?

"No, your Honor, but I read his book," Lovejoy responded.

"But did you talk with him?"

"Your Honor, I believe as sure as I'm standing here that when you read someone's book, you talk to them. I believe I talked to George Washington, and the signers of the Constitution and Henry David Thoreau. *Don't you talk to Oliver Wendell Holmes when you read his books*?"

Smith was impressed, but not swayed. He called a short recess and returned with a unique and somewhat bizarre ruling. Gofman could testify to the record, but not to the jury. If Lovejoy were found guilty, the case would go to the State Supreme Judicial Court *before* sentencing to determine the validity of the testimony.

So while the jurors played pinochle in the back room, Gofman delivered a scathing indictment of the nuclear power industry. Under Lovejoy's questioning he told the court that he had worked on the Manhattan Project and with Glenn Seaborg, a chairman of the Atomic Energy Commission. The AEC's lax standards on low-level radiation, he said, were a "license to commit murder." As many as thirty-two thousand additional cases of cancer, leukemia and birth defects would result if nuclear development continued under such standards.

A reactor meltdown, he continued, could destroy hundreds of thousand of lives and do billions of dollars worth of damage. An area the size of Pennsylvania would be made uninhabitable for centuries. Nuclear proponents had issued statements saying the chances of a meltdown were miniscule, Gofman said, but that begged the question: "I find when we're talking about a mass of a hundred tons of material at five-thousand degrees Farenheit, with water around there, with hydrogen being generated, burning explosively, melting through concrete into soil, when somebody tells me that 'we're sure it isn't going to go far away,' I look at them as a chemist and I say 'I've heard various forms of insanity, but hardly this form.'

"I don't really know whether the chance is one in ten, or one in a hundred, or one in ten thousand. I just ask myself in view of the fact that we have so much easier ways to generate energy needs, why do it this way?"

The brunt of Gofman's testimony centered on plutonium, on which he had done much pioneer research. Gofman told the court that, in the Atomic Energy Commission's phrase, plutonium is "the most fiendishly toxic substance ever known." Three tablespoonfuls, he said, could cause nine billion human cancers.

But each nuclear plant creates thousands of pounds of waste plutonium, and there's no way to store it. "The proliferation of nuclear power carries with it the obligation to guard the radioactive garbage...not only for our generation but for the next thousand or several thousand."

Gofman said plutonium has a half-life of 24,000 years and must be guarded "99.9999 percent perfectly, in peace and war, with human error and human malice, guerilla activities, psychotics, malfunction of equipment. . . . Do you believe there's anything you'd like to guarantee will be done 99.9999 percent perfectly for 100,000 years?''

Gofman capped his testimony with a conspiracy charge. ''Some awfully big interests invested in uranium and the future of atomic power,'' he said. ''And unfortunately their view is 'We've got to recover our investment, no matter what the cost to the public.' ''

The scholarly, bearded Gofman cut a striking figure on the stand, and his testimony had enormous impact on the local community. A reporter for the *Greenfield Recorder* later wrote a column admitting that Gofman's testimony had convinced him to rethink his stand on nuclear power.

Lovejoy's next witness was radical historian Howard Zinn, an expert on civil disobedience and an honored veteran of antiwar and antidraft cases. After Zinn's credentials had been established, Lovejoy asked him whether he thought the tower-toppling had been malicious. ''No,'' Zinn blurted out ingenuously. Even as prosecutor Murphy was leaping to his feet, the jurors began gathering their wraps.

Smith then let Zinn testify as Gofman had, without the jury present. Under questioning from Lovejoy, Zinn told the court that the tower-toppling was in the best tradition of Gandhi, Thoreau, and the Abolitionists, including (of course) Elijah P. Lovejoy, Sam's distant cousin, who was hanged by a proslavery mob in southern Illinois.

Judge Smith interrupted to ask if true civil disobedience didn't demand both strict nonviolence and the acceptance of lawful punishment.

Zinn replied that the destruction of property was not violent when life was at stake. ''Violence,'' he said, ''has to do with human beings, not property.''

Zinn pointed out that Lovejoy had turned himself in, whereas many civil disobedients disappear rather than stand trial.

Smith, who was looking and acting more like Spencer Tracy every day, seemed much taken with Zinn, and constantly interrupted him with questions. A good third of what Zinn said was in the form of colloquy with the judge. At one point Smith asked leave for a private conversation with the witness, and leaned over to talk quietly. Zinn said later the judge had asked to meet him for dinner sometime.

When Zinn finished, Lovejoy called a few character witnesses, as well as one Bruce Olmstead, an "environmental engineer." He testified that NU had sold him the wrecked tower for $250 and that he had used it to make three windmills.

Lovejoy Takes the Stand

Finally, Lovejoy took the stand himself. The jury of nine women and five men, freed at last from their backroom confinement, were all ears. Northampton attorney Thomas Lesser did the questioning.

Lovejoy began by talking about growing up as an army brat, then, after his father was killed, living on a farm near Springfield. There, he said, an old Yankee farmer taught him to respect the balance of nature.

In high school he studied math and physics, but dropped out of Amherst College to work in the Springfield area.

Returning to Amherst, he was graduated in political science, and then had moved to Montague Farm.

His mind was made up about nuclear power during a trip to Seattle. There he read in local papers of a massive leak of radioactive wastes from a storage tank at Hanford, in eastern Washington state. More than one hundred thousand gallons had escaped from holding tanks into the ground. The incident had been concealed by the AEC and Atlantic-Richfield until some investigative reporters found out about it.

When the story was printed, the AEC had a comeback. A computer printout, they said, showed that the liquid wouldn't reach the Columbia River (thus destroying it) before the year AD 2700. Until then, the Commission said confidently, everything would be fine.

That, said Lovejoy, was it. For six months he read everything he could get his hands on about atomic energy, finally settling on *Poisoned Power* as the basic text. The more he read, Lovejoy told the jury, the more he was convinced nuclear power plants were "the most horrendous development our community has ever faced."

And the more he looked into legal recourse, the more the AEC seemed like "a kangaroo court...a panel that acts as promoter and regulator, judge, jury, and thief all rolled into one."

Returning to Montague, he saw the tower for the first time, and knew it would have to go down. He wasn't sure he'd be the one to do it, he said, but the tower definitely had to go.

Lovejoy talked to the jury for six hours about his life and conversion to sabotage, without objection from the prosecution. The last hour was an intensely emotional narration of his final decision to topple the tower, how he acted not out of malice "but because I had fallen in love with a little four-year-old girl named Sequoyah. I asked myself, who am I to do this thing, to take on the role of judge. But then I thought about this little girl who couldn't defend herself, and I knew I had to act."

After the trial several jurors said they were deeply moved by Lovejoy's testimony. A poll indicated a hung jury, at least eight to four or nine to three in Lovejoy's favor. Much would have depended on Judge Smith's directions, which might have been favorable on the malice question.

But it never got that far. There was another aspect to the indictment, and it read "destruction of personal property." Under Massachusetts law, destroying personal property is a felony punishable by five years in prison; destroying real property is a six month misdemeanor. The prosecution wanted Sam behind bars for five years, not just six months.

Smith had expressed doubt all along that the tower could pass as personal property. It was worth $42,500, nobody disputed that. But when Lovejoy produced two Montague tax officials who testified the tower had been assessed as real property, and when Murphy called an NU official who affirmed under cross-examination that the tax had been paid as real property, everyone knew it was all over.

So, after lunch on Yom Kippur Eve, Smith convened court, again with the jury sent out, and announced his decision. He was going to void the charge because he "could not in good conscience ask a jury to deliberate on an indictment with a hole in it."

Lovejoy practically begged him not to do it. He had meant the trial to test the issue of nuclear power, and he wanted his guilt or innocence to be determined on that issue, by the jury and "the people of Franklin County."

Smith replied with a lecture on the law. "Justice is justice is justice is justice," he concluded.

Then he called in the jury, ordered them to stand and render a verdict of "Not guilty." He then dismissed the court. The crowd was as stunned as the jurors were relieved.

A Corporate Meltdown?

The Lovejoy trial had an immense impact on the surrounding community. Everyone had an opinion. Its impact would undoubtedly have been far greater had Lovejoy been acquitted by the jury instead of the judge. But even at that the trial hammered into the mind of the Connecticut Valley the twin issues of civil disobedience and nuclear power.

And new developments were not long in coming.

Three days before Lovejoy's acquittal, the AEC ordered twenty one of America's fifty active nuclear plants shut for an emergency safety check. A reactor at Zion, Illinois, had sprung a leak in a cooling pipe, and a check of a similar reactor revealed a similar crack. The other nineteen reactors of that type—all made by General Electric—were ordered to close within sixty days. Six more were shut in Japan. It was the largest multiple shutdown in the history of atomic power.

That same day Carl Hocevar, a leading computer analyst at an Idaho company doing testing on reactor safety, quit his position in protest. The true dangers of nuclear power were being covered up, he said, and he wanted to be free to tell the truth. He then joined the Union of Concerned Scientists in Cambridge, Massachusetts.

One day after Lovejoy's acquittal, Northeast Utilities announced that the Montague project would be postponed for at least one year. The reason: money.

Company officials had begun harboring doubts about a capital investment of $1.5 billion, only a third of which they can count on. In August, 1974, NU President Lelan Sillin admitted "the company must raise $1 billion to build the Montague plant, and when $1 billion is needed, and when interest rates are as high as they are, we have to look seriously at the situation."

For nuclear opponents, the delay looked like a big step down the road to cancellation. Given the current rate of inflation, even one year's delay could put the plant cost up to $2 billion. Within the past year [1974] at least thirty projected nuclear plants have been postponed or canceled because of the money squeeze, and there seemed no reason why Montague should be an exception.

But despite its problems, NU had taken great pains to make clear that the project has not been canceled, and that political opposition will make "no difference" to its plans. Charles Bragg, NU vice-president for public relations, told the *Greenfield Recorder* at the very outset that local opposition "wouldn't affect us. We would have to go ahead with it even if there was a protest movement mounted by the citizens of the area."

☆☆☆

It soon became evident that ramming the Montague nuclear power plant into the Connecticut Valley might not be all that simple. In November, 1974, three weeks after Lovejoy's acquittal, the referendum against the Montague plant failed by the slim margin of 52.5 percent to 47.5 percent. Nearly 23,000 of roughly 48,000 voters in the Franklin-Hampshire-Hampden Counties state Senate District registered clear opposition to the project, despite relatively little exposure to the issue. No candidate running for any public office in the region openly supported the initiative.

Most of the local press expressed shock. The *Greenfield Recorder* columnist who had likened Lovejoy to Adolph Hitler now conceded that support for the project was "melting away."

The best indicator was Montague itself. In the spring of 1974 the community voted for the plant by nearly three to one. Six months later the ratio fell to less than two to one, with the anitnuclear vote swelling from 770 to 1091—a jump of better than forty percent.

The proposition asking for the dismantling of the Vermont Yankee and Yankee Rowe plants also failed to get a majority, but did far better than expected. Some of the candidates and most of the press viewed the proposal as an irresponsible joke at public expense, and predicted it would get 20 percent of the vote at best. Instead 15,313 people—33 percent of the voters— registered a desire to bury two active reactors valued at approximately $1 billion. Wendell, on Montague's northeast border, became the first town in America to produce a majority vote for dismantling, by 98–68.

Over the next few years, organized antinuclear activism spread throughout the Connecticut Valley. The Alternative

Energy Coalition became the *only* AEC when the federal Atomic Energy Commission was dismantled to make way for a more "responsible" regulatory agency, the Nuclear Regulatory Commission (NRC).

Meanwhile, Northeast Utilities continued to have money troubles, and was forced to ask Connecticut and Massachusetts ratepayers for a long string of rate hikes—with diminishing success. On February 22, 1975, one year after the toppling of the Montague tower, NU announced it was delaying construction for at least four years, for financial reasons. The company persisted, however, in asserting that the plant would be built.

Soon thereafter, Green Mountain Post, a film-makers' collective based in Turners Falls, issued a sixty-minute documentary entitled *Lovejoy's Nuclear War*, dealing with the Montague fight and the basic issues of atomic energy. The film became a grass-roots phenomenon, and was shown to church, community, and antinuclear groups around the country, as well as at high schools and colleges. Soon Green Mountain Post had sold more than 120 prints of the film, and estimated it had been seen by more than four million viewers.

Meanwhile, increasing numbers of scientific and consumer groups began expressing serious doubts about atomic energy. Bit by bit, the issue began to receive serious treatment in the national media. Then, in the fall of 1976, a nuclear bomb explosion in China brought home an ugly reminder of the horrors of the atomic age:

China Bomb Test Affects Valley Milk
(*Valley Advocate*, October 20 and 27, 1976)

CONNECTICUT VALLEY RESIDENTS got an unpleasant jolt when radiation was discovered in area milk last week.

Official sources in Boston discounted the dangers of drinking the contaminated milk, but politicians and environmental groups in Massachusetts, Connecticut, and New York expressed dismay and bitterness about both the levels of radiation and the government's response to the problem.

According to Gerald Parker of the state Radiation Control Department in Boston, the radiation came from a recent Chinese nuclear bomb test, and reached a peak here around October 10. Parker said the debris that has affected vast regions of the East Coast of the United States was in evidence in such a wide range of geographical locations that there was little doubt it came from the Chinese bomb. The levels of radioactive iodine 131 found here were in the range of 400–600 picocuries per liter of milk, roughly "in the same ball park" as the levels found in Connecticut and other New England states. "We don't start worrying about the radioactivity," he said, "until it gets in the range of 80,000 picocuries."

Parker added that the level of fallout from this test was in the range of only one-tenth that of the fallout from American and Russian testing done in 1961–62.

Under Parker's orders, the Massachusetts Farm Bureau ordered the state's commercial dairy farmers to bring their cows inside on Friday, October 8. This precaution was taken because fallout tends to sit on grazing land and be ingested by the cows and then secreted in milk. Hay stored in barns would not be contaminated. And since the iodine has a short half-life, it would be less dangerous in a few days, thus, presumably, making pasture land safe again. Parker said the Farm Bureau told the dairies they could put their cows back on pasture the following Monday.

They Weren't Told

But apparently a number of small dairies and individual home owners didn't get the word. Milk tested independently by Dr. K. S. R. Sastry at the University of Massachusetts showed radiation at 500–1000 picocuries per liter. Milk from one Montague farm showed a reading of 1600 picocuries.

Better Luck Next Time?

Federal Environmental Protection Agency (EPA) guidelines indicate that "protective action" should be taken when tests reveal an iodine level of 100 picocuries per liter.

Although test results for iodine 131 have been readily available, tests for strontium and cesium are not yet completed. Strontium and cesium have longer half-lives. While iodine tends to concentrate in the thyroid, strontium builds up in the bone cells. Parker indicated there were no preliminary indications of deadly plutonium, but that an element known as neptunium was apparently present in some quantity.

There was also considerable disagreement as to what "acceptable" levels might be. Parker originally referred to a July, 1964, Federal Radiation Control report citing 84,000 picocuries as the level at which one liter could be considered a dangerous dose.

According to Parker, EPA statistics on iodine 131 doses of the nature involved in the Chinese fallout might result in "less than two additional deaths in the next forty-five years."

But Jay Kastner, chief of the Environmental Standards Branch of the Nuclear Regulatory Commission, confirmed that federal standards call for action when milk reaches a level of 100–1000 picocuries per liter. "Beyond that," he said, "we think about ordering cows off of pasture, and withholding milk." Kastner said that the EPA, not the NRC, was responsible for notifying the states of potentially high levels of radiation.

He also confirmed that the NRC received notification from a number of nuclear power stations that there were irregularities in their readings during the fallout period, and that a number of reactor operators feared their instruments had registered abnormal radiation releases from the plants themselves.

On October 6, Senator Edward Kennedy issued a statement to the effect that he was "disturbed" by the fact that news of potential danger from the test "was not made public until ten days after it happened." As chairman of the Senate Subcommittee on Administrative Practices, Kennedy wrote a letter to a number of administrative agencies requesting a full accounting of their procedures for dealing with subsequent situations. Among other things, he cited "considerable disagreement" on the exact dosage of radiation that could result in hazard to public health.

Connecticut State Representative John Anderson lumped the federal response to the Chinese fallout together with a recent spillage of nuclear wastes being trucked from the Connecticut Millstone nuclear plant to Barnwell, South Carolina. The two experiences, said Anderson, "show how ill-equipped we are as a state and as a country to deal with nuclear problems. They [the NRC] want all the power, but they're taking little of the responsibility."

Friends of the Earth (FOE) also attacked the NRC, charging it "failed to alert the public or protect them against radioactive fallout." FOE is demanding to know why the Energy Resource and Development Administration and the NRC "failed to issue any public health alert to authorities or the public even though they had been tracking the fallout eastward across the U.S."

FOE also took issue with the government's standards on acceptable levels of radiation dosage, citing Dr. Ernest Sternglass of the University of Pittsburgh, who charged that at the 400–600 picocurie levels found in both Massachusetts and Connecticut, pregnant women drinking the milk for a two-month period could expect an increase of fetal deaths and leukemia on the order of 10 to 30 percent.

☆☆☆

The discovery of radiation in Connecticut Valley milk, and the official confusion that accompanied it, deeply frightened many area citizens. The tension was heightened because Vermont Yankee just dumped 83,000 gallons of radioactive tritium into the Connecticut River. The operators said the spillage was "accidental" and assured the public it posed "no danger." But health officials recommended that local residents avoid eating river fish for a while, and nuclear opponents charged that the tritium had been dumped on purpose to relieve Vermont Yankee's waste storage problem.

The spill and the Chinese fallout marked a turning point for many. Even three years after the toppling of the Montague tower, the nuclear issue had remained somewhat abstract.

But known radiation releases were now polluting their food and their river.

By the winter of 1975–76 the eyes of the nuclear opposition were turning north, to Seabrook, New Hampshire (next chapter).

But in the Connecticut Valley, Northeast Utilities persisted in its moves toward building the Montague plant. In the winter of 1976–77 NU issued its final environmental impact statement, and it bore the official NRC stamp of approval:

Bringing the War Back Home
(*Valley Advocate*, February 23, 1977)

NORTHEAST UTILITIES, THE NUCLEAR REGULATORY COMMISSION, and the powers that pull their strings have issued a Declaration of War against the living things of the Connecticut Valley.

The Declaration, published last week, comes in the form of a pithy baby-blue volume entitled *NUREG–0084*, or, *Final Environmental Statement, Montague Nuclear Power Stations Numbers One and Two*.

It is a criminal document.

It states, among other things, that the citizens of the Connecticut Valley can stand radiation, can stand the destruction of their river, can stand a lethal threat to their weather patterns and water supply, can stand twin 565-foot cooling towers in the midst of some of the world's loveliest terrain, can stand the devastation of the area economy, can stand the arrogant manipulations of a multibillion-dollar conglomerate supported by us but beyond our legal control, and can stand, ultimately, the erection of an unnecessary, unwanted, and immoral monument to human greed, deception, authoritarianism, and self-destruction.

What the report says, quite simply, is that Northeast Utilities can go ahead and build its 2300-megawatt nuclear power plant on the Connecticut River.

That's what it says. But whether NU can actually build it depends on one basic question: How hard are we willing to work to stop it?

An environmental impact statement is supposed to reflect the sincere and deep-seated concern on the part of our government—which is, after all, OUR government—for the health, safety, and natural well-being of the people in the area of a proposed nuclear power plant. Ostensibly, this statement should lay out the facts about the environmental impact of the Montague nuke.

What we've been given instead is an arbitrary, uncaring, deceitful coverup that has been aptly described by one local group as "shoddy, inaccurate, misleading, and incomplete." The report runs several hundred pages and has but one basic message: Northeast Utilities is hereby sanctioned by the United States government to do almost whatever it wants.

Perhaps the treatment of the plant's most visible feature—its cooling towers—can demonstrate the basic tenor of the report best of all.

In order to keep the radioactive core from melting through the reactor and into the Montague Plains, NU wants to suck in roughly 25 percent of the entire Connecticut River.

There are now some serious questions about whether there's actually enough water available. Among other things, it seems the Montague nuke, which is supposed to produce electricity, may force the inactivation of at least one hydroelectric dam.

The cooling towers themselves would be 565 feet tall—ten feet taller than the Washington Monument.

The spectre of twin tubular monsters in the midst of our Valley has been, for many of us, something of a joke. Too absurd even to contemplate.

But the time for laughing is over. Northeast Utilities is clinically insane. The towers would be visible for miles. The steam they would throw into the atmosphere (which would be chemically treated and thus contaminate crops throughout the Valley) could reach on a windless day nearly a mile into the air.

There is, ironically enough, a clause in this environmental statement where the NRC staff rules out the cooling towers as having too great an aesthetic impact. The clause is, I was informed by an NRC spokesperson, "a mistake," left in the report in error.

And so it goes. Towers appear, disappear, and reappear again with a stroke of the pen.

But the words in this report are anything but abstract. They represent blocks of concrete, steel girders, bulldozers, radioactive fuel, and giant chunks of our money inching ever closer to the giant dump they want to make of the Montague Plains.

Aside from threatening the health, safety, and environmental well-being of all of us, this plant would mean economic catastrophe for the

Valley. It would force the construction of new roads, raise taxes and utility rates in the towns surrounding Montague, bring a crisis in such public services as sewage and schools, and place the area economy on a boom-bust cycle that would once again screw the vast bulk of us for the benefit of the few. Good for the land swindlers, bad for the folks. Good for the people already rolling in bucks, catastrophic for everybody else.

The jobs being used as bait by the company would materialize only for a closed fraternity of well-paid workers who will flood the Valley from all over and then leave it to build the next nuke—unless, of course, NU decided to go for four or six or eight.

The fact is, we are being lied to. The government-approved report makes no real mention of the possible contamination of Quabbin Reservoir, neatly skirts the economic disruption issues and skims the question of cancer, leukemia, and birth defects so closely associated with radiation. We don't hear much about the lack of nuclear insurance for individual families, the lack of real electrical demand to justify this plant, or even the lack of a clear mandate on the part of the citizens of the Valley or the customers of Northeast Utilities to build the thing.

Apparently, neither the company nor the government gives a damn about what the neighbors think.

But we all have the right to protect our health, safety, and homes, even if the corporations think we don't.

The Montague Nuclear Power Station will never be built, basically because there are people in this Valley who possess the proverbial fire in the belly.

The publication of this report marks a turning point in the progress of the plant, and it must also mark a watershed for the opposition. Those of you who have been on the fence, those of you who have been waiting until things got a little hotter—the time has come. The fences are down. The heat is on.

There is no escaping this issue. Those of you (and there seem to be quite a few) who are considering fleeing the Valley to escape the plant—forget it. If nuclear power can't be stopped in the Connecticut Valley, it can't be stopped. This is among the most powerful, best-organized grass-roots antinuclear movements in the world. If the industry can plough up the people and build reactors in this Valley, they can build them anywhere. If the people are to win the battle against the nukes, it will be won in New England, here, at Seabrook, at Vernon, at Plymouth, at Charlestown, and at Millstone.

Embodied in the nuclear issue is virtually everything we must confront in American society. We know natural energy will create more jobs for the general population than nukes—and that these jobs will be accessible to all levels of society, regardless of previous training.

We know that the fuel companies and the utilities are lying about how much energy we have and how much we need—and that conservation and solar energy will wreck their financial monopoly as well as their political control over our power sources.

We know radiation, hot water, and steam will have a disastrous effect on our bodies and our environment—and that putting health and the environment before corporate profits runs counter to the basic assumptions of the system as we now know it.

Like this farce of an environmental impact statement, all that is old business.

What's new is the coming spring, the approaching public hearings at Montague, and the do-or-die necessity of building a solid, working, broad-based movement that can support successful opposition both here and at Seabrook.

There is no more holding back.

There is no escaping involvement,

The war has come home.

3
Seabrook 1976: The Shot Heard Round the World

☆☆☆

Nuclear War by the Sea
(*The Nation*, September 11, 1976)

SEABROOK, NEW HAMPSHIRE, is no college town or bastion of the middle class. It is a tight-knit community of 5700, proud of its insular New England heritage, proud also of its lovely beaches and long, breath-taking marshes that host thousands of magnificent sea birds and a universe of marine life.

In 1969 the Public Service Company (PSNH), which supplies New Hampshire with 90 percent of its electricity, announced its intention to build twin atomic reactors in the midst of those marshes. The plan was so little publicized that one Seabrook resident, Tony Santasucci, didn't even know the company wanted his land until he sprained his ankle in one of their drill holes. He still walks with a limp, and with most of his neighbors he vows to fight to the bitter end. "I've been around sixty-two years," he says, "and I'm too old to move. We don't need that plant and if you ask me, Public Service is a bunch of liars. They'll never kick me out of here. They'll have to drag me out first."

Most of Seabrook shares the sentiment. Despite extravagant promises of tax benefits, jobs, and business, the town voted against the plant in March, 1976, by a margin of 768 to 632. By June the town opposition had joined with environmental groups from around New England to form the Clamshell Alliance, an umbrella coalition of fifteen separate antinuclear organizations. The goal of the Alliance is nothing less than to do whatever is necessary within a nonviolent framework to occupy the plant site, halt construction and force the PSNH to cancel the project.

Loeb and Thomson

They face formidable opposition. New Hampshire is the domain of William Loeb, publisher of the *Manchester Union-Leader*, and his side-kick Meldrim Thomson, the Governor. They make as reactionary a ruling pair as any state can point to, and they are both ardent nuclear promoters.

The Loeb-Thomson team has been stung deeply by the defeat of Aristotle Onassis's multimillion-dollar oil refinery at the hands of the citizens of Durham, and more recently by the town of Walpole's rejection of a major pulp mill.

During the Onassis and Walpole debates, Loeb and Thomson were trapped by their own archconservative "home rule" rhetoric. But the nuke is too big a deal to permit mere ideology to interfere. Blaming the plant's defeat at a Seabrook town meeting on "outsiders," Thomson encouraged Public Service to ignore the vote, a request with which the PSNH gladly complied. Thomson then backed it up with an order to state employees that they either keep their mouths shut about negative feelings toward the project, or else lose their jobs. If a state employee felt compelled to question the plant, he said, then that employee should "resign his state job and go out and oppose it."

The Governor was soon forced to rescind the order, especially in light of the fact that his own Attorney General, David H. Souter, joined the Attorney General of Massachusetts in filing an exception against the federal Nuclear Regulatory Commission's (NRC) grant to the PSNH of a construction permit. Among other things, the two Attorneys General claimed that the final Environmental Protection Agency impact studies had not yet been completed.

This month—August, 1976—a moratorium on licensing new plants went into effect. But until then the NRC had never denied a nuclear construction permit. At Seabrook, however, Ernest Salo, one of the three members of the Atomic Safety and Licensing Board, cast a rare minority vote against the Seabrook permit, claiming that the plant would have a destructive impact on commercial and recreational fishing in the area.

His concern is widely shared. A key feature of the PSNH plan is a pair of water pipes each fully nineteen feet in diameter, designed to suck in and spew out ocean water for reactor cooling at the rate of more than 800,000 gallons per minute. Thus more than a billion gallons a day would be returned to the sea fully forty degrees hotter than normal. Local fishing people and ecologists claim this could have a devastating effect on the offshore marine environment.

Money Problems

Strong questions are also being raised about the economics of the plan. The PSNH generating capacity is already 1343 megawatts, more than twice its average hourly sales for the past two years, and more than 40 percent higher than the state's peak electrical demand. As originally planned, PSNH half-ownership in twin 1150-megawatt reactors would nearly double the company's capacity.

But, as with many other utilities, the PSNH's sales growth has dropped drastically, rising only 0.1 percent in 1974 and actually dropping 0.5 percent in 1975. Such figures have thrown the company's need for the additional capacity into serious question. In fact, the general downswing in electrical demand growth that has forced cancellation of nuclear plants across the United States has already pushed the company into a tenuous financial position at Seabrook. Last December, President Lelan Sillin of Northeast Utilities cited a drop in sales and pulled his company out of its 12 percent commitment to the Seabrook reactors. Two other utilities followed suit, leaving the PSNH responsible for a 70 percent interest rather than the 50 percent originally scheduled, and raising serious doubts about whether the company really has the assets to finance the project.

Cost estimates aren't proceeding exactly on course either. Four years ago the company said the plant would cost $973 million. Today it puts the figure at $1.6 billion, but few outside the PSNH believe it. David Lessels, finance director of the state Public Utilities Commission (PUC), recently published a thirty-page report putting the cost in the vicinity of $2.53 billion and going up. The current $900 million underestimate, he said, could eventually cost the company's 250,000 customers a startling $3,600 each.

Here Come the Clams

Such figures are common to nuclear construction across America. What is uncommon about Seabrook is the nature of the resistance. "This movement is built from the bottom up," says Anna Gyorgy, a Clamshell volunteer from Montague, Massachusetts, where Northeast Utilities is planning twin reactors. "Here the movement starts with the town. There is no other way."

With a solid local base to work from, the Alliance is trying to turn Seabrook into a genuine regional face-off. Indeed, winning the confrontation is crucial to the nuclear industry for precisely that reason. With the costs of nuclear construction soaring, and with resistance to it growing in the country, it would seem that New England is the most promising area left for nuclear economics. It is devoid of coal and oil deposits, relatively low on hydroelectric power, and its long winters and heavy urbanization make it the area in America where nuclear power seems surest to be financially desirable. There are now seven nuclear plants online in the six New England states, with four more at various stages of construction and at least four more on the drawing boards.

But there are also strong antinuclear campaigns growing throughout the region, and they are beginning to coalesce at Seabrook. Many feel that if a major project can be defeated on the turf of New England's most reactionary state government, the impact on nukes in New England can well be fatal, particularly if the conflict is well-organized, well-publicized, and drawn out, all of which Seabrook promises to be.

There follows a basic question: Will there be enough political push, enough organization, and enough physical bodies to prevent nuclear construction? And with a five-year lead time before the plant is scheduled to open, how much time between will be required to stop it in its tracks?

To answer that, the Clamshell has adopted Gandhian nonviolent tactics. The first skirmish was a one-man show staged by Ron Rieck, a twenty-two-year-old apple picker from Weare, New Hampshire. On a bitter cold January 4, Rieck scaled a 175-foot PSNH weather tower at the site. Equipped with warm clothing, a little food, and two sheets of plywood, Rieck camped on the tower for thirty-six hours before the cold forced him down. Upon reaching earth he found Seabrook Police Chief Louis Promise waiting for him with a thermos of hot tea. The Company balked at charging Rieck with trespass, and he was later acquitted of one count of "creating a public disturbance."

Rieck's mini-occupation was followed on April 10 by a rally of three hundred people at the site, which was soon thereafter bulldozed and closed to the public.

On August 1, six hundred nuclear opponents, representing every state in New England, arrived gathering in the marshes. The rally of speeches, food, and song served as a send-off for eighteen Clamshell occupiers, all of them New Hampshire residents, who set off for the plant site. Carrying a score of pine and maple saplings, and accompanied by some forty journalists, they confronted PSNH officials with their wish to beautify the area. When the company offered a compromise plot of ground, the "Clams" refused it and started gardening where the PSNH wants to plant reactors.

Police then informed both the demonstrators and the media that they were all under arrest. The order was quickly countermanded for the news teams, but the Clams sat down next to their saplings and allowed themselves to be dragged through two hundred yards of mud and underbrush to waiting police vans. At the Hampton Falls police station (Seabrook's is too small for such a crowd) the eighteen were booked on charges of criminal trespass, disturbing the peace, and resisting arrest. (Later, they were all convicted on one charge or the other.)

Four days later the PSNH and Governor Thomson arrived in full force for the official ground-breaking ceremonies. The date was set for August 5 so the news would break in the papers on August 6, Hiroshima Day, apparently the PSNH's idea of a joke.

The game was spoiled, however, by a dozen Seabrook residents, many of them elderly, who sat in chairs in the middle of the road through which the official procession was to travel to the site. They were removed roughly, but not arrested, despite Meldrim Thomson's vindictive oratory.

The actual ground-breaking ceremony (accomplished with silver-plated shovels) was free from disruption. But the extremely tense atmosphere was hardly relaxed by a long day's program of guerrilla theatre, scattered vehicle blockages, and three arrests, including that of Guy Chichester, forty-one, a local carpenter at whose house the Clamshell had been formed a month earlier. Also arrested was Gretchen Siegler, twenty-four, who managed to infiltrate the official luncheon at the posh Exeter Inn in order to disrupt Thomson's speech.

Coincidentally, Presidential candidate Jimmy Carter had arrived in nearby Manchester to celebrate the New Hampshire primary victory that launched his successful campaign for the nomination. Cornered by Green Mountain Post Films, Carter's reply to questions on civil disobedience and nuclear power made national news.

"I've always felt," Carter said, "that anybody who disagrees with the civil law in a matter of conscience has a right openly to express that disobedience. At the same time, under our societal structure, it's necessary that they be willing to take the consequences of their disobedience.

"I believe that there's a place for nuclear power in the future. It ought to be minimized, it ought to be a last resort; [there] ought to be tough safety precautions guaranteed by the President and other leaders in Washington, with nuclear power plants located where people don't live, where the environment will not be destroyed, with the reactor core beneath ground level, where the reactor building is tightly sealed and they have a standardization of design, to make sure the people can have confidence in the safety of nuclear power plants. . . ."

On August 13 the NRC announced a temporary moratorium on issuing new licenses. The decision stemmed from a court ruling forced in part by the New England Coalition on Nuclear Pollution, which was taking legal action against the reactor at Vernon, Vermont. The moratorium does not affect Seabrook, which already has its license, and the NRC move was greeted here with a good deal of skepticism. "They'll hold their standard few days of hearings," says Chichester, "and then it's back to business as usual. It's nothing to be excited about."

In any case, the Seabrook opposition is waiting for neither Jimmy Carter nor the NRC. On August 22, some 1,500 people attended a rally at the Hampton Falls common, near the site. 180 of them marched to the site and onto it. They were arrested without violence or injury and held overnight in the Portsmouth Armory, about twenty miles away.

Most of the Clam occupiers were charged with criminal trespass and released on personal recognizance. But ten were charged with contempt of a Superior Court injunction and immediately railroaded off to jail . . ..

Trial of the Seabrook Ten
(*Valley Advocate*, September 8 and 15, 1976)

THE RAILROAD WAS SET IN MOTION on August 20, when Maurice Bois, a justice of the Superior Court of Rockingham County, was approached by the PSNH to issue an injunction against the upcoming occupation. Judge Bois had been approached for a similar injunction prior to the August 1 occupation, but had refused.

This time he did not refuse. There were headlines about it, but only a few copies of the injunction had been delivered to the antinuke office at Rye, New Hampshire. Some twenty people were named in the injunction, including all of those arrested August 1. No more than five actually received written copies of the injunction.

Because the injunction came down on Friday afternoon, just hours before the occupation, there was no chance for an appeal. Attorney Thomas Lesser made a hurried trip to Judge Bois on Saturday morning, but Bois refused to lift the injunction.

The way the law reads, it may not matter whether the injunction is later proven to be invalid or unconstitutional—it must be obeyed and tested later. A person can be convicted of violating an injunction and be forced to serve the full sentence even though the injunction is later overturned.

There is also another catch: if the judge agrees not to hand down a sentence of more than six months, he can try you himself, without a jury.

Thus it happened that ten people who were arrested on both August 1 and August 22 were charged with criminal contempt of the injunction, and were forced to stand trial before none other than Judge Maurice Bois, the man who issued the injunction.

What made the situation even more bleak was that the judge had been appointed by New Hampshire's pronuclear Governor Meldrim Thomson not once, but twice. A Manchester lawyer, Bois was widely known as a "Loeb man" who had done work for the *Union-Leader*. Thomson had appointed Bois to the Superior Court in the first place, and then had appointed him to move up to the state Supreme Court within a month after the trials began.

Another Trial

On August 30, when pretrial motions began, Judge Bois came on strong. As ordered by Bois, Lesser appeared in court on Tuesday. He attempted to withdraw, and the eight *pro se* (self- defended) occupiers attempted to dismiss him so they could get local counsel (two of the defendants, Brian Cullen and Ann Carol Riley, had a local attorney hired by their parents). Bois refused to hear of Lesser's dismissal, and made him a virtual prisoner of the court by ordering him to remain at least in the role of legal adviser.

Bois also denied motions that he withdraw because of prejudice, or that he delay the trial so that adequate defense could be prepared. Bois did allow the defendants to cross-examine prosecution witnesses, but then refused to allow any questions concerning the validity of the injunction itself.

The judge then conducted a game of psychic jujitsu with the defendants, browbeating them whenever possible. He persisted in calling Neil Linsky by a variety of names, ranging from "Lipsky" to "Lishky" and got into a tiff with defendant Medora Hamilton over her wish to be addressed as "Ms."

Much to everyone's surprise, however, the judge did allow a motion to view, which meant that after lunch on Tuesday the court proceeded to the site of the plant. On a gorgeous, hot, clear day, Bois was guided through the now devastated marshland, inspecting the August 22 battleground. "Your Honor will please note," said defendant Jay Adams, "that this once was a beautiful place to live."

Revelations

Over the course of the next three days the defendants found their balance in the courtroom. With Lesser sitting at their back, the eight *pro se* occupiers grew accustomed to the dual role of attorney and defendant.

In the meantime, Bois appeared to mellow. In the course of the pretrial hearing, he had laid into the record a mind-boggling number of seeming contradictions, hostile statements, and other potential errors that could be grounds for later reversal. When he was hit with a defense motion to remove the trial to federal court, it might well have occurred to him that the affair was going to travel beyond his courtroom. Judge Hugh Bounes, sitting on the federal bench in Concord, refused to take on the case before it was completed at the state level. But he denied the motion "without prejudice," meaning he was interested and might well be willing to hear the case later.

With that over his head, Bois attempted to clean up the record a bit. He allowed the eight to cross-examine prosecution witnesses extensively, resulting in a number of interesting revelations. Seabrook Project Manager John Herrin revealed that he had called the NRC immediately following the occupation, and Governor Thomson had personally visited the site thereafter to "congratulate" the officers in charge. Deputy Sheriff Lelan Davis revealed that he had been specifically ordered *not* to deliver copies of the injunction to everyone named on it. And State Police Colonel Paul Doyon told the court that he had discussed enforcement of the injunction with the state attorney general *before* Bois signed it.

At 11:30 A.M. Thursday, the prosecution rested its case. By 3:45 P.M. Friday, the defense had also finished. Each of the eight made final statements, tying their defense against contempt to their horror of the dangers of nuclear power. Kevin Hopkins briefly differentiated contempt of court from contempt for a nuclear power plant, while Michael Cushing delivered a long attack on the governor and the legislature, concluding that the injunction violated his civil liberties. Mary Gregory told the judge that "ignorance of the law is no excuse, but ignorance of the dangers of nuclear power is also no excuse. All my actions have sprung from concern for my children and my children's children."

Every evening the ten were prepared to go to jail, and each evening they were surprised that the judge himself had prolonged the trial.

Finally, after a three-day weekend, Bois reconvened court and ended the suspense. All the defendants were guilty as charged. He was sentencing them to six months in prison, three suspended, and he was denying them bail pending appeal. Officers of the court then carried off the ten defendants, and Bois, satisfied, left the courtroom.

☆☆☆

After six days in jail, the Seabrook ten were freed by the New Hampshire Supreme Court. Bois had also thrown Robert Gross, a New Hampshire attorney, into prison for his role as mediator at the site even though Gross had not been arrested there. The multiple convictions were generally viewed as a direct assault on civil rights in New Hampshire, and prompted U.S. Senator John Durkin (D.-N.H.) to charge that Bois's conduct had made the state look like "the Mississippi of the North," a comment that should have offended the people of Mississippi.

The trial of the Seabrook Ten made it clear, however, that the stakes were going to be very high in New Hampshire, and that somebody was taking these protests very seriously.

The Clamshell at the time was gearing up for another occupation at the Seabrook site, to take place on October 23. But soon after calling the action the Alliance decided to reverse itself, and to stage a legal rally and energy fair at a nearby state park instead. Among other things, the Bois trials had convinced some of the Clams that the organization was a little thin to be undertaking yet another mass illegal action. Others also felt that a legal energy fair might attract some of the more conservative citizenry of the seacoast, and could furthermore be used to demonstrate some of the machinery of alternative energy technology.

After much internal wrangling and recrimination, a whole run of posters was scrapped and the Alliance's "Natural Energy Fair" attracted some three thousand people to Hampton Beach State Park—directly across Seabrook Harbor from the nuke site—on a blustery, cold weekend.

The Clams then set April 30, 1977, as the date for the next occupation at Seabrook, and dug in for a long winter's organizing. Results of the national election indicated a long struggle ahead:

New Battles Loom in Nuclear Controversy
(*Valley Advocate*, November 17, 1976)

ALONGSIDE THE VICTORY OF JIMMY CARTER, the big nuke news in the November 2 national election was the crushing defeat of six referenda aimed at regulating nuclear power. Reformers in Oregon, Washington, Colorado, Montana, Arizona, and Ohio attempted to control nuclear expansion by such means as legislative review, insurance limitations, and other measures.

The propositions were generally rewrites of the California Proposition 15, which failed two-to-one last summer. This fall's crop of initiatives differed in that they exempted existing plants from proposed regulations, but they were identical in that they were all defeated, in most cases by equally overwhelming margins.

"The results are dismaying and perplexing," says Herb Epstein, of Critical Mass, an organization formed by Ralph Nader. "We expected to win at least two or three states. We're really going to have to re-evaluate this thing."

Epstein said polls showed the initiatives with substantial leads in Oregon, Washington, and Colorado only three weeks before the voting. But the final vote split 58–42 in Oregon; two to one in Washington and 72–28 in Colorado. "In all cases we were outspent by a huge margin," he complained. "The utilities advertising made it much more difficult for the public to understand the initiatives, which finished by saying 'If you're confused, vote no.' "

Sandy Sterrett of Ohioans for Utility Reform said pronuke forces there so confused the issue that "by the time they were through the voters didn't really know what a yes or a no vote meant."

Sterrett said her group spent approximately $27,000, while the utilities poured in something between $1.3 and $2 million. According to Epstein, the ratio in the other states was similar, with antinuke forces being outspent by a factor of as much as one hundred. "We've got to figure out how to deal with this money thing," said Epstein. "And we're also looking into coming up with simpler, one-issue referenda questions. In the future we'll probably concentrate on just one issue, like economics or waste."

No Pu in Mich., No CWIP in Mo.

In fact, two such issues did pass this fall.

A ballot on nuclear wastes provided a sharp defeat for the nuclear industry in three extremely conservative rural counties in western Michigan. The counties—Alpena, Oscata, and Presqu'Isle—are in the northwestern corner of the state, where geological features form what many scientists feel would be an ideal dumping ground for America's fast-growing stores of radioactive wastes. "We first heard something was up when rumors started coming out of the state Department of Natural Resources," explains Bill Morey of the Alpena County Citizens for Participation in Natural Energy Decisions. "They said the feds had come around wanting a decision on some preliminary drilling. Later we found out what they really wanted to do here, and a lot of people were angry."

Morey said the government sent in two scientists to assure the citizenry of the safety and benefits of a terminal storage facility, but that the people "resented the way the feds were trying to bust in."

Morey said the government was "vague" about the job benefits of the project, and that stories of problems with a storage facility at Maxey Flats, Kentucky, began filtering through the area. The anti-dump forces spent $100 to the utilities' $2500, but won the vote in all three counties by an overwhelming nine-to-one margin. The final count in Alpena County was 10,132 to 1,214.

A state-wide vote in Missouri also brought success to utility opponents. The contest centered on a billing procedure known as Construction Work in Progress (CWIP), which allows utility companies to make customers pay for generating facilities before they are completed.

Kay Drey, a chief initiative organizer, said the campaign was not meant to be antinuclear, but the utilities portrayed it as such. "Even out-of-state utilities sent letters to their Missouri stockholders claiming that an end to CWIP would mean an end to nuclear power," she said.

By November 2, Drey says, Missouri utilities had already spent at least $350,000—and perhaps as much as $1 million—to defeat the anti-CWIP initiative. "All the vice-presidents and their families were out at the polling places passing out literature," Drey says. "I think they were really afraid, because the voters are very unhappy about their electric rates."

The final score was a stunning victory for the reformers, 1,109,123 to 673,432.

Congressional Changes

...Meanwhile, Dixie Lee Ray, former chairperson of the Atomic Energy Commission and an ardent nuclear promoter, won the governorship of the state of Washington. With her victory and the defeat of the referendum there, nuclear opponents in that state may face a rocky road.

But in Congress, a committee change is brewing that may cut the other way. For some time now, the Joint Committee on Atomic Energy (JCAE) has been a powerful pronuclear arm, forcing through legislation favorable to the industry and generally blocking floor debate with its ample procedural leverage.

But this fall two of its leading Senate members retired—Stuart Symington (D.-Mo.) and John Pastore (D.-R.I.). And three more of its members—James Buckley (R.-N.Y.), John Tunney (D.-Calif.) and Joseph Montoya (D.-N. Mex.)—all failed in re-election bids. The loss of five members—plus the campaign of antinuke Senator Mike Gravel (D.-Alaska) to gain a seat on the committee—may help bring the body's function to an end. There has already been a move to abolish the committee, and the new Carter administration may help. "If that happens," says Critical Mass's Epstein, "you can expect a whole new atmosphere in Congress toward nuclear power. The JCAE was a very strong roadblock to giving the antinuclear side a hearing."

☆☆☆

The Joint Committee on Atomic Energy was, in fact, soon abolished. But while the 1976 election provided mixed results for the antinuclear movement, there seemed little doubt about the feelings of the Nuclear Regulatory Commission. On November 5, three days after the vote, the NRC ended its temporary moratorium on nuclear licensing. The abrupt cutoff, coming so quickly after the election, seemed to confirm critics' suspicions that the moratorium was called in the first place to cool the nuclear issue during the national campaign.

On November 8, however, the PSNH got an unexpected jolt from the Environmental Protection Agency. John McGlennon, head of the regional EPA office in Boston, withdrew the agency's temporary approval of the Seabrook cooling system. "The company has failed," he explained, "to demonstrate that the proposed heated water discharges will assure the protection of fish and wildlife in and on the affected coastal waters." The decision, complained company officials, came "like a thunderbolt out of the blue." The action, said PSNH public relations man Frank Shants, "is arbitrary, fickle, and capricious. It defies description. We just don't understand how a decision like that could have been made."

PSNH President William Tallman echoed the sentiment. "I'm appalled at the decision, and on behalf of every man, woman, and child that we serve, we promise we will take every single legal step possible to overturn the decision."

The company's response stemmed from the fact that without EPA approval, the NRC's grant of a construction permit was technically void. But the commissioners decided that even without the permit, PSNH could continue limited construction at the Seabrook site pending an appeal to the national EPA. It was the second time the NRC had exempted Seabrook from a construction halt in the five months since construction had begun (the first exemption came in August, when the Atomic Safety and Licensing Board ruled Seabrook should be included in the licensing moratorium, and was quickly overruled by the top commissioners).

This second reprieve did little to help the PSNH's finances. By fall of 1976 the company claimed to have poured some $100 million into the Seabrook site. Now, while construction proceeded, it was unclear what would happen to that investment. The PSNH had appealed McGlennon's ruling to the national EPA. If they lost, either a new cooling system would have to be designed, or the project would have to be scrapped.

Either way, the company wanted a quick decision.

But national EPA director Russell Train refused to rule, deferring instead to Jimmy Carter's incoming, yet-to-be-named appointee.

Train's refusal put the PSNH in a tough spot with investors and banks from whom it needed loans to continue construction, and with its partner utilities, who began to pull out. In January, the company announced a 90 percent cutback in the construction force at the site, slashing the project budget for January-June, 1977, from $124 million to $16.6 million. Within six months after starting construction, the Seabrook plant seemed on the brink of bankruptcy:

Seabrook Nuke Funds Melted Down
(*Valley Advocate*, January 19, 1977)

THE PSNH IS A SMALL COMPANY as utilities go, with something on the order of a half-billion dollars in assets. It does not have the kind of cash to throw around that is normally required to build a nuke. In fact, one of the strongest criticisms leveled by New Hampshire opponents of the plant was that the company couldn't really afford it, and was getting in way over its head. As the lead utility on the project, the PSNH is committed to 50 percent of the twin 2300-megawatt plant.

At the time of the original planning for the plant, the cost was estimated at $900 million. But rising material, labor, and other costs, plus delays, have caused the current projected cost of the plant to soar to $2 billion, nearly 250 percent the original estimate. According to David Lessels, a member of the New Hampshire Public Utilities Commission, the final cost could well be more like $2.6 billion, and opponents of the plant claim that if it is completed it will most likely ring in at more than $3 billion.

The soaring costs and cost projections have added to an air of investor uncertainty. On December 23, 1975, Northeast Utilities clouded the atmosphere by announcing its desire to sell off its 12 percent share in the project. NU at the same time sold off a 13 percent share in the proposed Pilgrim II nuclear plant, and downgraded its ownership in Millstone III at Waterford, on Long Island Sound.

In announcing the decision, NU President Lelan Sillin cited a downswing in electrical demand and a desire to improve NU's cash-flow situation.

But nearly thirteen months later, NU is still holding its 12 percent share in Seabrook, having been unable to find a definite buyer. The New England Power Company (NEPCo), which services areas in Massachusetts, New Hampshire, and Rhode Island, has been slated to take up the bulk of NU's shares. But NEPCo's Paul Hartrey confirmed Friday [January 14, 1977] that the company is holding back. NEPCo already owns a 10 percent interest in Seabrook, but, says Hartrey, "we don't think it is wise to invest an extra $300 million in a project where the cooling system is in the air."

Connecticut's United Illuminating (UI) would also like to drop its investment from 20 percent to 10 percent. UI spokesperson John Betts attributed the decision to a drop in demand. "We don't need 20 percent of that plant for our needs in the eighties," he explained. "We're not even back to 1973 demand yet."

But UI may have a problem getting rid of its Seabrook stock. NEPCo was going to buy a piece, but is now holding off. UI was, according to Betts, also going to sell off a 1.1865 percent share to the Central Maine Power Company. But Central Maine now says they'll also be holding off.

Nuclear Confusion

What it all adds up to is a severely confused financial climate, now made chaotic by the EPA ruling. "As long as the uncertainty hangs over the project," says Tallman, "it would not be possible for us to do any permanent financing. The cash matter has a bearing on our thinking. Not only are we trapped in a vise between two conflicting administrative agency decisions, but we are caught in a chasm of changing administrations."

EPA Administrator John Quarles admitted that the Seabrook decision would in fact have a major impact on the approval of all future cooling systems for water-cooled plants, nuclear or otherwise. And in the face of that, the PSNH seems confident that the EPA will go their way. "There are about 250 applications for variances just like ours now pending," said PSNH public relations man Norm Cullerot. "Turning down our system would have a severe impact nationally. We just don't think that will happen."

☆☆☆

By spring the town of Seabrook had voted to "ban transportation or storage of nuclear materials associated with power plants," and eight other towns bordering Seabrook had gone on record opposing the plant or supporting the town's opposition.

In Vermont, thirty-four town meetings voted to "exclude construction and operation of commercial nuclear reactors or any other nuclear facility and the transport, storage, or disposal of radioactive wastes for such reactors in land, air, or water of the town."

At roughly the same time, the Central Maine Power Company abandoned its plans to build a nuclear reactor at Sears Island, opting for coal instead. It was the first clear cancellation of a nuclear project in New England.

Construction at Seabrook continued to limp ahead. But the nuclear opposition mushroomed. Clamshell organizers pushed their battered cars around the region, preparing for the upcoming April 30 occupation. What they came up with won the antinuclear movement world-wide attention.

Meanwhile, the man who had come to New Hampshire as the Democratic candidate for President just after the first Seabrook arrests, and who had labeled atomic energy a "last resort," became America's chief executive:

Carter's Choice – And Ours
(*New Age*, January, 1977)

AFTER EIGHT YEARS of the lowest common denominator, the environmental movement now faces a President who solicited its support and who clearly understands the basic issues and language of the natural ecology.

Environmentalism is no abstract issue. Cancer is now a disease of epidemic proportions in the United States. Part of its cause may well lie with the fallout from the massive atmospheric testing carried out by the United States and the Soviet Union in the early sixties. Pollution is another major factor—carcinogens let loose in the form of oil spills, dirty air, radioactivity from power plants, and additives in our food.

Opposition to these plagues is not simply a question of aesthetics; it's one of survival, of the health, safety, and genetic future of the human race.

This phase of the environmental movement was born under Richard Nixon and Gerald Ford. It has grown expecting that at every step it would be confronted with the most outrageous lies from straight-faced corporate executives and labor leaders, and that those lies would be backed up by the highest office in the land. It has sweated blood for even the most marginal legislative victories, only to have most of them vetoed by the executive branch. It won its first major national battle—over construction of the SST—by a single congressional vote, despite all-out opposition from the White House.

Now we have Jimmy Carter. There's a little-known episode in Carter's life which makes his election especially intriguing. In 1952 Carter was senior crew officer of the U.S.S. *Sea Wolf*, a nuclear submarine being built under the auspices of Admiral Hyman Rickover. In August of that year, the experimental NRX atomic reactor at Chalk River, Canada, experienced a partial meltdown which turned the core into a twisted wreck. The Canadians asked for help, and Carter was one of a number of officers dispatched to help clean up. As part of the procedure, technicians were forced to run into the mess, do some menial task (such as removing a bolt), and then run out before the radioactivity knocked them dead. They wore gas masks and as much heavy shielding as they could carry and still move. But the area was so "hot" that they usually had to be in and out in two minutes or less.

Carter participated in one of those crews, getting his full year's radiation dose in eighty-nine seconds. In his autobiography he mentions that he and his crew returned home joking about the preferability of "death versus sterility."

One would like to think that any person participating in such an event would emerge with an abhorrence of atomic energy. As Ralph Nader put it, if Carter fails to act on nuclear power, "It won't be because he doesn't know the danger; it won't be because he doesn't have the knowledge; it won't be because he doesn't have the authority. It will be because he doesn't have the guts."

It is now quite clear that atomic energy is entirely unnecessary for meeting America's energy needs. Studies by Amory Lovins (published in the November issue of *Foreign Affairs*, of all places) and others clearly demonstrate that nuclear construction can stop and the plants in existence be phased out in a very short time, if the slack is taken up by fossil fuels for the decade or two it will take to put solar, wind, tidal, and geothermal energy—in concert with a sweeping program of recycling and conservation—into total control.

The United States, representing 6 percent of the world's population, consumes more than 30 percent of its energy resources. The per capita consumption here is twice that of most countries in Europe, and almost unimaginably higher than that of the countries of the Third World. We have been taught to waste energy; the economy is built on it.

If Carter is held to his campaign rhetoric, his inauguration should be a major landmark in the decline of nuclear power. "Unless we conserve energy dramatically," he said many times during the campaign, "make the shift to coal, and substantially increase our use of solar energy, we will have no alternative to greatly increased dependence on nuclear power. As one who is intimately familiar with the problems and potential of nuclear energy, I believe we must make every effort to keep that dependence to a minimum."

Nukes are hardly the only polluters on the planet, but they are number one in terms of both long-range damage and potential profits —as much as $1 trillion or more by the year 2000. If Carter is willing to take on the nuclear industry, he might also be ready to take on kepone, PCBs, recombinant DNA, and a few other nightmares.

But is it wise to expect anything from a President of the United States?

Woodrow Wilson won re-election in November, 1916, with the slogan "He Kept Us Out of War!" By April, 1917, America was at war.

Franklin Roosevelt pledged in the fall of 1940 that no American soldier would set foot overseas while he was in office. By December, 1941, American troops were in foreign trenches.

Lyndon Johnson professed to be a peace candidate in 1964, then escalated the war in Vietnam.

Richard Nixon "had a plan" to end that war in 1968. Four years later he was sending the heaviest bombing missions in human history over Vietnam, carrying out the most comprehensive, and conscious, environmental destruction campaign ever executed.

So now comes Jimmy Carter, an intriguing engineer with environmental support, conservationist campaign rhetoric, and the demonstrated moral and intellectual capacity to understand the issue, something that was never clear with Johnson, Nixon, or Ford.

What if he does come through?

What if he acts as an environmentalist President?

What if he does move against the corporations?

If we learned anything from Vietnam, it was that meaningful, lasting change can only come from the bottom up. Nothing really moves in society until the people as a whole are convinced that it should.

If we really expect to build a system in the United States—and the world—whereby the economic and industrial life is conducted in harmony with the natural environment, then we are talking about a movement which will affect every woman, man, and child on the planet. We're talking in terms not of four or eight years, but four or eight decades.

The role of James Earl Carter in the next few years will be fascinating to follow. He might be a closet environmentalist crazy for all we know. But even if he does everything he possibly can in the interests of the ecology, he'll still have to stand in line with the rest of those doing educational work. No bill he might sign, no reform he might push, will have any lasting value until it's fully grasped, digested, and accepted by the general population.

Nothing will change in this country—not on the issue of war, or racism, or the economy, or the environment—until the people *want* it to change, until we all really *want* a spiritual and material order based on equity with our bodies, our planet, ourselves.

In that kind of long-term campaign, the President of the United States, no matter who he or she is, is just another heart, gently pounding.

4

Seabrook 1977:
The Battle Escalates

☆☆☆

High Tension in the Energy Debate–
The Clamshell Response
(*The Nation*, June 18, 1977)

Seabrook, N.H.

ON APRIL 30, more than 2,000 marchers descended on the nuclear construction site here. Thanks to their numbers, their discipline, coolness and determination—but thanks also to the ill-considered response of the state's highest authorities—this demonstration by aroused citizens will almost certainly be remembered as a watershed event in the direct-action politics of the 1970s.

By the middle of the week of April 23, the alliance had announced that it could count on a least 1,000 occupiers, and that it had trained at least 1,800. While the organization raced to solidify last-minute arrangements, the realization that something big was about to happen mobilized the media and began attracting scores of additional occupiers.

The news also attracted the frustration and wrath of the nuclear plant's chief backers, New Hampshire's archconservative Gov. Meldrim Thomson, and his patron, William Loeb.

Together, Thomson and Loeb used the week before the occupation to create an atmosphere absent from this country since the days of Vietnam. Labeling alliance members "Communists" and "perverts," the Governor and the newspaper charged that, despite the group's public commitment to nonviolence, the occupation was being used as a "cover for terrorism." The real intent of the action, they said, was to bring on bloodshed, and they soon produced an "inside report" from the ultrarightest U.S. Labor Party to "prove" their case.

Meanwhile, a crucial behind-the-scenes struggle was going on between Thomson and State Police Commander Col. Paul Doyon. Doyon, who had handled the August 22 arrests, was given a full account of the plans for April 30 from the Clamshell Alliance. Much to Loeb and Thomson's chagrin, he apparently decided it would be wiser to allow the occupiers onto the site and arrest them there than to try blocking them at the edge of the property. A confrontation at the plant's border would have snarled traffic on Route 1 for hours, if not days, and might have degenerated into real ugliness.

It soon became apparent that the shouting by Thomson and Loeb was aimed as much at Doyon as at the Clamshell Alliance. Then, on Thursday, two days before the occupation, Thomson met with alliance members Cathy Wolff and Robin Read. "I think the meeting blew the governor's mind," said Read afterward. "He suddenly had to confront the fact that we are also human beings, and that we were, in fact, committed to nonviolence."

By the clear, pleasant morning of April 30, there remained hardly a trace of the hysteria that had dominated the state's headlines for a week. Several hundred new Clamshell recruits arrived for final training sessions, and there was no counterdemonstration or police resistance in sight.

Occupiers had been arriving from Maine, New Hampshire, Massachusetts, and elsewhere in the Northeast, many of them camping overnight. The bulk of them were in three campgrounds to the west of the plant, and they came together at the main access road, used each morning by the site's laborers. By 3 P.M. this "western route" contingent numbered more than 1,300; accompanied by some 200 members of the press, they marched without incident down the half-mile asphalt road and onto the site.

When all the groups had converged on the hot, dusty parking lot at the belly of the site, the occupation force numbered well over 2,000. Singing, chanting, and in the best of spirits, the occupiers unpacked their knapsacks and tents, organized themselves into neighborhoods, dug latrines, and settled in for the night.

While most of the occupiers went about the business of living, a more or less permanent conclave of elected representatives took root at the southwest corner of the campsite. Among other things, this decision-making body passed ordinances against the construction of nuclear power plants or the transportation of radioactive materials within "town" limits. It also sent messages of solidarity to the workers and environmentalists of the world, and to the 3,000-person rally being held across the marsh at the Hampton Beach State Park by the Concerned Citizens of Seabrook and Hampton Falls, an organization of local people opposed to the plant.

At about 2 P.M. on May 1, Colonel Doyon and Governor Thomson appeared at the east side of the parking lot to meet with six chosen representatives of Occupation City and ask that they vacate the premises. Colonel Doyon offered the use of school buses chartered by the state to facilitate the departure.

Elizabeth Boardman, a long-time associate of the American Friends Service Committee, which has played a crucial role in the growth of the Clamshell, told the Colonel, "We are all mutually sorry. But our purpose was not simply to draw attention to the occupation and capture media attention but to stop all construction of the proposed nuclear power plant."

Doyon replied that he, too, was "very sorry," and at 3:09 read a notice over a series of police loudspeakers that everyone who did not leave the site would be arrested. A body of New Hampshire, Maine, Vermont, Connecticut, and Rhode Island State Troopers then emerged from nearby warehouses to begin the arrests. Massachusetts Gov. Michael Dukakis was the only New England chief executive to refuse Thomson's request for police aid. He was, he explained, "unconvinced" that the Seabrook occupation posed a significant threat to the security of the region.

The arrests proceeded smoothly, if slowly. Piling people onto chartered school buses and into National Guard troop carriers, the patrolmen took more than twelve hours to haul everyone away. Some of the occupiers who had packed up their gear in the afternoon unpacked it again and reopened their tents to wait through the long hours of the night. Many were not arrested until dawn.

Meanwhile, a major bottleneck had been created at the Portsmouth Armory, where those arrested were being processed. The first batch was booked and released on personal recognizance.

But suddenly, late in the afternoon, Thomson swooped in by helicopter. Hampton District Court clerk Mac Hamilton later testified that he had come to the armory with a box full of personal recognizance forms, but that a change of procedure went into effect with the arrival of the Governor. Now the rule was $1,500 bail, which nearly all the prisoners were unwilling or unable to meet.

Three District Court judges soon arrived at the armory, a number of "guilty" pleas resulted in some occupiers being hauled off to prison. At least two contempt charges were leveled in the course of the night. "It was a travesty," said attorney Emmanuel Krasner of Rochester, New Hampshire, who had been engaged by the Clamshell to represent those occupying the site, and who had been present most of the night. "The state violated just about every basic right in the Constitution."

When the sun finally set on May 2, some 1,400 detainees were being held under makeshift conditions in four National Guard armories around the state. Some had spent up to fifteen hours in buses and in the backs of the National Guard troop carriers, as well as in closed, stuffy semitrailers while awaiting arraignment.

Many of the prisoners were held for three days and more, without beds, telephone access, or more than perfunctory legal counsel. Friends, family, and the external Clamshell support system were unable to learn who had been arrested, where they were being held, or what the conditions of confinement were like. "It was a classic police-state situation," said a Clamshell support-worker. "It was like 1,400 had been 'disappeared' by the state."

Conflicting reports soon trickled out as to conditions in the armories. Nearly seven hundred occupiers were being held in the Manchester Armory alone, and overcrowding seemed a problem at all four. There were reports of communicable disease, of one outbreak of food poisoning, of difficult, if not impossible, access to telephones, and of irregular access to outsiders. The state remained uncooperative throughout the entire two-week detainment, and a complete accounting of who was being held where was finally obtained only after great difficulty.

Both the alliance organization and the body of occupiers were caught unprepared for the extended detention, the group expectation having been for immediate release on personal recognizance. But while a steady stream of occupiers bailed out, it rapidly became clear that a very large core was more than willing to remain "indefinitely." "Many of us felt that the bail system is unjust to begin with," said Cathy Wolff, who was held in Somersworth. "The process favors those with money. We also felt bail was unjust in this case because it was being used as a punishment instead of insurance that we would appear for trial. We weren't happy about costing the taxpayers all that money, but we figured we'd be saving everybody money if what we did helped stop the nuke."

The detention hit a legal low point on Thursday, May 5, when the first of the prisoners were brought out for their District Court trials. Murray Rosenblith, twenty-six, a Brooklynite and staff member of the radical pacifist *WIN* magazine, was brought along with sixteen other defendants from Somersworth to stand trial before Judge Alfred Casassa at Hampton District Court.

As expected, Casassa found Rosenblith guilty of criminal trespass, and then sentenced him to fifteen days at hard labor and a $200 fine, with the jail sentence suspended pending appeal on $200 bond.

The verdict and sentence both seemed routine. Therefore it was a shock when, after lunch break, state Attorney General David Souter, swooped up to the court in a limousine and appealed to Casassa to raise the ante. Speaking for half an hour, Souter told Casassa that "giving the defendants a suspended sentence is tantamount to no punishment at all." The Superior Court, where the cases would be appealed, was already running behind schedule. "Most likely," Souter complained, "these cases will never clear the docket there." As a result, "justice can only be done by imposing sentence right now." As a guarantee, he added, no appeal bail should be allowed; sentences should begin to be served immediately.

Defense attorney Krasner responded by pointing out that not one of the August 22 defendants had defaulted on his or her personal recognizance bond, that the occupiers clearly intended to stand trial, and that the refusal of bail or imposition of high bail was tantamount to punitive detention. Most of the Superior Court trials would be scheduled weeks—even months—in the future, far beyond the fifteen-day period set out in the sentences. To impose the penalties now, or to set high or no bail, Krasner argued, was to deny the defendants their right to appeal.

Casassa, apparently persuaded by Souter, imposed unsuspended sentences on the next sixteen defendants, and demanded a $500 bond. He then called Rosenblith back into the courtroom, lifted the suspension on his sentence and upped his bail. "It was quite amazing," said the pacifist writer. "At Portsmouth we were told we would be released on personal recognizance. Then it turned into $100. This morning it was $200, and this afternoon it's $500. What next?"

By Sunday, May 8, the number of occupiers remaining in the armories had dropped below 1,000, as several hundred who had not yet been called to trial availed themselves of the chance to bail out at the original $100 mark. The Governor announced, "We are winning the battle of Seabrook." His optimism was premature. For one thing, it had become clear that the prolonged detention gave the Seabrook occupation moral weight and dramatic impact it would never have carried as a mere weekend confrontation. For another, blame for the arraignment foul-up and the expense of imprisonment (it was costing up to $50,000 per day) was being laid at the Governor's doorstep. Local newspapers, as well as the *Boston Globe* and *The New York Times*, noting the implications of Souter's appearance at the Hampton court, editorialized for the prisoners' release.

By Monday, May 9, ACLU lawyers Nancy Gertner and John Reinstein were in U.S. District Court at Concord, in front of Judge Hugh Bounes, demanding immediate emptying of the armories because of the physical conditions there. The suit challenged the use of bail for punitive purposes, and included a class-action damage suit demanding $5,000 per arrest and $5,000 per detainee per day. Based on the irregularities of the detentions, the suit drew heavily on precedent set by actions following the May 1, 1970 arrests of more than 13,000 antiwar demonstrators in Washington, D.C. A preliminary ruling (now on appeal) granted more than $1 million in awards in that case. The 1977 Seabrook suit, which grew by as much as $5 million per day, would soon rise well above $50 million.

As the second week of detention dragged on, a nerve-racking chess game developed between the Clamshell and the Statehouse, with the National Guard, the court system, and the taxpayers holding decisive sway. Inside the armories, the occupiers established a cordial and fruitful relationship with the Guard. "We treated them like fellow human beings," said Rennie Cushing, a Clamshell organizer detained in the Manchester Armory. "A lot of the Guard were against the nuke to begin with, and a lot more were against it by the time the occupation was over. I think we broke new ground, and if I were Governor Thomson I'd think twice about relying on them to keep us off that site next time." Personal relationships reached a point, in fact, where Gertner and Reinstein subpoenaed two Guardsmen to federal court to serve as character witnesses for the occupiers.

Meanwhile, as it became evident that a large segment of the occupiers had no intention of bailing out, the cost of detaining them became a political hot potato. The Rockingham County commissioners announced in mid-week that they would refuse to pay costs of imprisoning convicted occupiers, and that they would sue the Governor if the county were ordered to pay for the armory detentions.

Finally, negotiations began between Rockingham County Prosecutor Carlton Eldridge and the alliance. After a series of meetings, an agreement was reached whereby Eldridge undertook to ask the judges in his district to grant personal recognizance if the occupiers would accept mass trials in return. It would, after all, be the Rockingham system that would bear the burden of trying the Seabrook 1,414. And with public outcry mounting, Eldridge seemed eager to settle.

On Thursday, May 12, he and alliance negotiators signed an accord that was quickly approved by the detainees at the Somersworth and Dover Armories, as well as by the inmates at Portsmouth, which had been reopened for prisoners earlier in the week.

At Manchester and Concord, however, serious objections arose, and at their insistence negotiations were reopened. Finally, at 5:30 A.M. on Friday, an elaborate compromise was reached which included provisions for a few individual trials, at least one public jury trial, refund arrangements for those who had already bailed out, and special conditions for noncooperators. "It was a nerve-racking head-to-head," says Charles Light, one of the key negotiators, "but I guess the weight of all those people was just too much for the state to carry."

Thus at 9:30 A.M. the first busload of occupiers arrived at the Hampton District Court, and a process began that would carry well into the night and would ultimately result in the release on personal recognizance of the remaining 541 detainees.

As the third Seabrook occupation thus drew to a media-saturated close, nuclear opponents found they had made a stunning impression on the world environmental scene. The occupation and prolonged detention had focused international attention on the nuclear issue as perhaps nothing else could have done, and had driven home the point that thousands of citizens were now willing to face arrest and imprisonment in order to call a halt to atomic reactor construction.

If the occupation proved the antinuclear movement had reached a new level of maturity and mass appeal, it also seemed a powerful testament to the tactics of nonviolence. For the third time the Clamshell Alliance had staged a mass civil disobedience action without a single incidence of violence or serious bodily harm. The tactics of peaceful action had opened the gates to the site when any other approach seemed certain to have kept them closed. It also maintained for the occupation an overwhelming base of credibility and popular support against which the Thomson administration was simply unable to respond.

Now, in the wake of its third tenfold increase in size, the alliance faces a critical period. Direct-action environmentalism has clearly accelerated from a small assembly of local groups to a full-scale movement, and with that must inevitably come all the growing pains of factionalism and organizational strain.

Meanwhile, the Seabrook plant continues to stagger along under regulatory, financial, and political difficulties. The Environmental Protection Agency has not yet decided whether it will approve the cooling system for the plant (which could be disastrous for the shellfish of the area); and until it does the banks will not advance the money needed to go forward with construction. More and more it appears a project doomed to an early death. If somehow it survives, Clamshell and the growing direct-action antinuclear movement around the country are almost certain to make Seabrook the critical battleground of nuclear energy. The April 30 occupation was, after all, accompanied by antinuclear actions at Browns Ferry, Alabama; San Luis Obispo, California; Portland, Oregon; Rocky Flats, Colorado; Wintersburg, Arizona; Zion, Illinois; Fulton, Missouri; Pittsburgh and Three Mile Island, Pennsylvania; and at Montpelier, Vermont. In their wake has come the Abalone Alliance, a statewide antinuclear group in California; and the Oyster Shell Alliance, aimed at a nuke under construction and another planned in the Mississippi Delta region of Louisiana. A Great Plains Alliance has been active in Missouri for some time, and it seems probable that other groups dedicated to direct action against nuclear power will be popping up everywhere.

If there should be a fourth occupation at Seabrook, it might well become a mass civil-disobedience action the like of which this country has not seen for quite some time. Full-scale occupations have already occurred in Europe, where 28,000 Swiss, French, and West German citizens occupied a nuclear construction site at Wyhl, West Germany, and where more than 20,000 French nuclear opponents tried the same thing at Malville, near Lyons. The former resulted in an on-site occupation lasting more than a year, and was ended only by cancellation of the plant; the latter presaged a widespread upheaval that has raised questions about the future of nuclear energy in France. Mass antinuclear opposition has also moved to civil disobedience in Italy, Switzerland, and Japan; it is threatening in Sweden, Spain, and Australia.

What comes next in the United States will depend in large part on the Carter administration, and the depth of its commitment to what it has termed "the last resort" in the energy crisis.

But what is now clear from the grass roots of New England is that the social movement which has developed on the issue has chosen a "last resort" of its own, and that movement is unlikely to slow down until nuclear power plants become no more than a bad memory.

☆☆☆

Although the spectacular confrontation between Meldrim Thomson and the Clams grabbed the world spotlight, a more subtle battle had already been won in the town. A rock-ribbed, conservative Republican community, Seabrook had made mass civil disobedience possible. The nuclear issue, once believed by both the industry and moderate environmentalists to be beyond the grasp of the average citizen, had saturated the grass roots:

The Opening Battles of the Eighties
(*Mother Jones*, August, 1977)

"I CAN READ!" confirms a boisterous Chris Peters from behind the counter. "And I can read between the lines: radiation has no conscience, no allegiance, no country!"

A jovial, vibrant man in his early fifties, Peters holds court at the cash register of his beer and cigarette outlet on New Hampshire Route 1, just north of the Massachusetts border.

Aside from owning a package store and a ballroom, Chris Peters does not like nuclear power plants. On April 30, when more than 2,000 fellow opponents marched onto the construction site of twin atomic reactors in Seabrook. New Hampshire, Peters was instrumental in their success. He had lent them not only his political oratory but also the parking lot of his Salisbury, Massachusetts, Marigold Ballroom as a staging area.

More than 700 occupiers had gathered there before marching the two miles across the state line, past Peters's package store and onto the access road that brought them and 200 representatives of the world media into direct contact with the nuclear industry.

We Won't Be Fooled Again

As a steady stream of customers flows through his carry-out for quick purchases of Schlitz, Kools, and an occasional bag of potato chips, Chris Peters makes political pronouncements.

"Nuclear energy should be disallowed," he proclaims in an eloquent cross between a song and a shout. "It is not cheap and is extremely dangerous. Any possible leakage of radioactive material would render the entire coastline obsolete."

While he rolls spectacular syntax out from among the Luckies, Peters flashes a shrewd half-smile, confirming the theatrical sense of an obvious master. The real key to the stunning success of the April 30 Seabrook occupation was that Chris Peters and a majority of the New Hampshire seacoast population not only supported it, they *loved* it. "Eighteen thousand next time?" he bellows. "Make it 180,000!"

For a solid week prior to the occupation, the *Manchester Union-Leader*—New Hampshire's most powerful paper—screamed that "hippies," "communists," and "perverts" were invading the state to foment revolution and promote expensive energy. The Governor backed it up with "inside information" purporting terrorism, bloodshed, lice, and rampant vegetarianism. It was a petty, sickening display.

But residents of the New Hampshire seacoast were not to be fooled. They provided the occupiers with crucial staging areas, gathered a barnload of food and created an atmosphere of appreciation and support. They neutralized the local police. They lined the streets to cheer. They flashed signs: "Seabrook Voted No Nukes" and "We Live Here—And Are Scared!"

While Americans, for the most part, may still favor atomic power plants, there's a very different feeling growing among those slated to live next door to them. It's a feeling created by too many nuclear mishaps, too many pushy corporations—and, apparently, a waning faith in the once-sacred Commandment that scientists know everything.

"That's right," says Tony Santasucci. "They got plenty of guys that are supposed to be geniuses. Well, they may be on paper. But when it comes to doing it, they're not so great. They got plenty of these plants that just don't work. If these plants are so safe, how come I can't get any insurance against nuclear accidents for my house?"

At sixty-nine, Tony has become something of a folk hero in the saga of the Seabrook nuke. White-haired, stocky, and powerfully built, Tony lives with his wife, Louisa, on five acres of land bordering the southern edge of the nuke site.

Tony had fought in World War II and was a truck mechanic in South Boston until he and Louisa retired to the tranquil New Hampshire seacoast thirteen years ago. The Santasuccis hosted a large group of occupiers—with about one hundred tents—on April 29, the night before they marched onto the site.

On their way to the Santasuccis', the Clamshell occupiers scaled an eight-foot wall of blasted rock that now rings the north and east side of Tony and Louisa's land. The pile was built there by the Public Service Company of New Hampshire (PSNH), which wants the Santasuccis' property. According to Louisa, a front company for Public Service had a real-estate agent threaten to bulldoze their house if they wouldn't sell. "You bring that bulldozer in," Louisa told him, "and I'll sit right in front of it." The real-estate man didn't come back. The Santasuccis still own their land.

If Not Seabrook, Then Where?

"You've got to realize how conservative New Hampshire really is," says Rennie Cushing, twenty-five, a Granite State native and one of three Clamshell Alliance members arrested at all three occupations staged at the Seabrook site. "There are no unions here with any real force. We had some of the biggest strikes in the country, but too many of them lost. They pay slave wages in this state."

Cushing raps lovingly and not without pain about his native seacoast. "The war was never here," he says. "The only time I saw it come home was once in Hampton, when the class of '68 football captain arrived from Quang Tri in a casket. People around here had their cars firebombed for antiwar stickers. A lot of people—including a lot who are against the nuke—are more than your average Republicans. They're militant conservatives. Personally, I think it's the winters."

It's also the *Manchester Union-Leader* and a thoroughly colonized economy. The *Union-Leader*, the only statewide daily, is the property of wealthy newspaper magnate William Loeb, who posts almost daily front-page editorials on pornography, world communism, and the wonders of atomic energy and nineteenth-century capitalism.

Loeb's money and presses elected Governor Thomson, Pittsburgh-born and Georgia-raised, a classic Third World mandarin. Thomson's basic livelihood has been the Equity Publishing Company, which does legal texts for, among other places, the goverments of Washington, D.C., and Costa Rica. Immediately after his re-election last November, Thomson flew to Costa Rica and wasn't heard from for a week. He took control of the Concord Statehouse by pledging "no taxes," either sales or income, a promise he's kept (so far) by virtually bankrupting the state while slashing nearly every state service in reach.

A staunch authoritarian, Thomson has advocated that the New Hampshire National Guard be equipped with nuclear weapons, and there are those who argue that the real reason he wants the Seabrook nukes is so he can build his own bomb.

At the very least, he and Loeb want the plant as an economic weapon. The state of New Hampshire already has much more electric power than it can use and, by most estimates, more than it will need for at least thirty years. To help rectify that, over the past couple of decades the PSNH, with the approval of the Public Utilities Commission, has ordered about twenty hydroelectric facilities into mothballs. At least one has been dynamited into oblivion.

The problem with the hydro plants was that they didn't expand the rate base of the PSNH, a private monopoly that supplies some 90 percent of the state's electricity. Like many major New Hampshire industries, the PSNH is a classic imperial venture. Eight of its top ten listed stockholders are from out of state, including the largest.

The power generated by the Seabrook nuke also would belong to out-of-staters. Located just over the Massachusetts border, the twin 1150-megawatt reactors were designed to send juice across state lines—at an expanded profit—to Boston, Worcester, Springfield, Hartford, and New Haven. The only way the plant really affects New Hampshire is to use the state's ground, its source of cooling water, and its political climate.

With the Statehouse, the *Union-Leader* and a long tradition of conservatism, New Hampshire must have seemed ideal for two, if not ten, nukes. And with high unemployment; low income; a marginal fishing, clamming, and tourist industry; and a history of rockribbed Republicanism, Seabrook must have seemed the ultimate nuclear target.

So the company came on with what has become a standard dance— glowing offers of high-paying jobs, expanded business opportunities, lower property taxes, and cheaper electric rates. For those who didn't buy the package, there were traditional threats of land seizure by eminent domain and warnings that the plant would be built even if the town opposed it.

"They treated us like peasants," says Rennie Cushing.

Roots of Resistance

Roots of the opposition can be traced to an oil refinery Aristotle Onassis tried to build in the seacoast town of Durham in the early seventies. The proposed complex stirred wide-ranging fears of destruction of some of the state's most scenic terrain and its fishing industry. (The danger to sea life also has been cited by Clamshell protesters, who point out that the nuclear plant's cooling system will raise the temperature of nearby ocean waters some forty degrees, killing off thousands of fish.)

After months of bruising political struggle, the refinery was canceled. There came to the seacoast the feeling, as Cushing puts it, "that it is possible to win." Shortly after Durham's victory over Onassis, attention shifted south to Seabrook.

"Nobody knew at first that they were going to build nuclear," says

Carline Peruse. "But then we read a letter to the editor in the paper from Guy Chichester. He knew a lot about it. And Tony called him up, and one thing led to another."

Peruse is in her sixties, a spry grandmother who has lived on Railroad Avenue all her life and can trace her Seabrook ancestry back eight generations. Large chunks of her neighborhood have been wiped away by the PSNH. For three years now she's been instrumental in the Concerned Citizens of Seabrook, a group of local residents who formed in hopes of learning more about the plant. They have now become staunchly antinuclear.

"The people of the seacoast are like people everywhere," says Guy Chichester. "They know when their lives are being threatened. All it takes is a little access to information." Chichester, a forty-one-year-old carpenter, brought his family from Long Island to Rye eight years ago. Chichester had participated in the Onassis oil fight and then became involved in Seabrook through the Seacoast Anti-Pollution League (SAPL).

SAPL's main focus was one shared by moderate antinuclear groups all over America: legal intervention within the Nuclear Regulatory Commission. "Our time and energy were going into the hearings presentations," Guy says. "And what we thought was that our little lawyer there, who everyone was going around scraping up bucks for, that he'd be able to do it.

"But on the right side of him there was a bank of lawyers that were getting $1,000 a day. And on the left side of him was a bank of lawyers that were getting $900 a day. It was a total gang-up picture."

As the hearings dragged on, information did begin to spread about the plant. Dolly Weinhold, a local opponent, made herself so expert in the geological make-up of the seacoast that she was able to conduct long cross-examinations at the hearings, earning herself the nickname "Earthquake Dolly." She single-handedly gave popular credibility to the fact that the Seabrook site is seismographically suspect.

Meanwhile, Chichester and others shifted their efforts from intervention to education and organization. The Concerned Citizens, which grew up through the energies of local biology teacher Cathy Foote-Silver, among others, began holding seminars and quiet get-togethers to raise important questions about health, safety and the environment. As the word spread, the opposition snowballed.

"Yeah, it was the nuclear issue," says Rennie Cushing, "but it was other things, too—the marsh, local politics, home rule, the clambeds. You can't separate any of it."

Even the PSNH's economic carrots began to lose appeal as the cost of the project skyrocketed from $900 million to $2 billion, and doubt set in about its local benefits.

"All this plant is for," says Tony Santasucci, "is nothing but a quick buck. They're not gonna lower the taxes here; they'll raise 'em. Because, with people coming in to operate the plant, they'll need more schools, more fire department, more police department; they'll have to put sewers in . . . . That's all gonna come out of the taxpayers of Seabrook. And the people of Seabrook don't have that money to spend."

Cracking the Egg

Gradually the antinuclear gospel became as firmly rooted in Seabrook as the granite. In March of 1976, the town meeting voted 768–632 against the plant.

In March of 1977 the vote was roughly the same to ban transportation of nuclear materials through the town, a gesture that in a democratic society should have called a halt to the PSNH's nuclear fantasies.

The second Seabrook vote was accompanied by votes of support from town meetings at Hampton, Hampton Falls, North Hampton, Exeter, Kensington, Durham, and Rye. Later, Salisbury, Massachusetts, which forms Seabrook's southern border, also checked in—ringing the site with towns in opposition.

"The egg is cracking," says Shirley Gustavson, a Concerned Citizen of Hampton Falls. A bright, sunny woman in her early forties, Gustavson worked for Republican candidates through 1972. "So many people thought nukes were clean and safe," she says. "Now they know better. And they're not so damned afraid of civil disobedience anymore. Of breaking the law."

"Look at *this*!" booms Chris Peters at the package store to a hapless pronuclear customer. "Look at these towers!" he yells, brandishing a *Union-Leader* sketch of the PSNH's latest scheme for cooling the nuke—two huge 590-foot towers. The customer winces, grabs his six-pack and scurries out of Peters' store.

Undaunted, Peters continues his harangue: "Can you imagine these monsters on our beautiful seacoast, throwing salt and steam over everything? It DOESN'T FIT! It just DOESN'T FIT!"

☆☆☆

The April-May occupation had an electrifying effect on the growing nuclear opposition around the world. In Europe, however, the movement was already far advanced. In June 150,000 Spanish nuclear opponents marched against a reactor planned for the Basque region. In July more than 30,000 French citizens marched against the Phénix fast-breeder project planned for the town of Malville. That march witnessed the first death of the French campaign, when Vital Michalon, a thirty-one-year-old schoolteacher, was struck and killed by a police tear-gas grenade.

In the United States, antinuclear alliances modeled after the Clamshell sprang up virtually wherever there was a nuclear mining, reactor, waste storage, or weapons site. On August 6, 1977, more than 120 demonstrations, rallies, and occupations took place around the country. Most of them focused on the releasing of hydrogen-filled balloons, with postcards attached, warning the downwind recipients that radiation could follow the same path. Accompanied by the birth of a new national disarmament group, the Mobilization for Survival, the August 6 actions were an unprecedented spontaneous uprising that signified far more to come. By fall mass nonviolent arrests had taken place at nuclear sites in California, Oregon, and Vermont, with more in planning stages.

But even as the Seabrook occupation was in progress, the beat of nuclear industry went on. On May 5, while Judge Alfred Casassa was upping the sentences on the Clamshell inmates, the Atomic Safety and Licensing Board okayed Tennessee Valley Authority construction of a 5100-megawatt nuclear facility—the world's largest—at Huntsville, Tennessee.

Meanwhile, the Carter administration seemed to be taking a slow but clear pronuclear bent. The President's appointments to the NRC included Peter Bradford, a Maine lawyer with a solid environmental background. But they also included Joseph Hendrie and Kent Hanson, both known to be staunchly pronuclear. Hanson's nomination was killed by Congressional opposition, but Hendrie slipped through to become the chairman of the NRC.

Carter's new appointment to head the Environmental Protection Agency was Douglas Costle of Connecticut, who had received favorable ratings from local environmentalists. Costle's first decision would be on the Seabrook cooling system. He waffled until some of the dust from the April 30 occupation had settled. And then on June 17, 1977, he went the wrong way:

The Lyndon Johnson of the Seventies
(*Valley Advocate*, June 29, 1977)

IN CASE YOU HAVEN'T GUESSED, the White House decision to approve the cooling system of the Seabrook Nuclear Power Station was a lot more than just a single decision on a single nuclear power plant.

Among other things it signalled the beginning of what will undoubtedly be the biggest domestic confrontation over any single issue in American politics since Vietnam.

Indeed, the Carter administration has thrown down the gauntlet on nuclear power as much as the Johnson administration did it on Vietnam in 1965, except back then the thud wasn't really noticeable for a couple of years.

This time, everybody heard.

If we listen back for a moment, we can hear Jimmy Carter campaigning in Manchester, New Hampshire on August 3, 1976.

Eighteen New Hampshirites had just been arrested at the Seabrook site—the first mass arrest in the U.S. over nuclear power—and Carter was approached by Green Mountain Post Films to learn the Democratic candidate's position. He replied that nuclear power would be a "last resort," that the plants should be built underground and their use "minimized."

Carter's statement on the need for precautions and others like it led a wide range of environmentalists to support his candidacy for President. It was the widespread popular belief—actively encouraged by the Carter camp—that his Presidency would minimize nuclear construction, whereas a Ford Administration would mean two reactors in every garage.

Yet on June 17, Douglas Costle, the Carter appointee to the Environmental Protection Agency, gave the green light for the Seabrook plant to be built. His decision overruled that of a Nixon appointee, John McGlennon, who as Regional EPA Director had turned down the plant. McGlennon himself reacted to Costle's decision by stating his disagreement, emphasizing that "there are many unknown risks in operating a cooling system that is going to discharge 1.2 billion gallons of heated water a day into a highly fragile ecosystem."

McGlennon's displeasure was confirmed by Drs. Edward Carpenter and Theodore Smayda, two professors who had served as EPA consultants on the decision. "Some of the conclusions based on the data were forced," said Smayda. "Others were contradicted by the data. That data was not sufficient to justify construction of Seabrook."

Nor does the Seabrook decision stand in a vacuum. Just prior to it, Interior Secretary Cecil Andrus paved the way for a nuke to be built at the Indiana Dunes on Lake Michigan, an environmentally unique area that had been protected by the Ford administration, which turned down nuclear construction there.

Indeed, if the whole thing brings on a remarkable *déjà vu*, take a sample of how Costle's EPA decision was announced. The ruling had been hanging fire since March, when EPA spokespeople said it would be "any day." The week of June 6, rumblings again came from Washington, but on Saturday the 11th the Associated Press reported that the decision was delayed for the weekend "in the wake of a review by President Carter."

"Carter wanted to review it himself," said an EPA spokesperson, "and I guess he found some questions he wanted answered. There's no doubt that Carter is concerned about what's in the ruling."

On Monday morning, the *Washington Post* carried a headline story that Costle had approved the cooling system. The EPA, however, refused to confirm or deny it, and Costle was not available for comment. He had not been quoted directly in the *Post* as saying the decision was go, only that it had been made. The *Post* had said it was a go-ahead, but there was no direct quote from Costle to that effect.

On Tuesday, rumors seeped out of Washington that there was severe infighting inside the EPA, and on Wednesday EPA officials said a decision had not, in fact, been made.

On Friday, Costle announced he had approved the cooling system "on narrow legal grounds" and insisted his decision was not a blanket go-ahead for nuclear power. He was promptly greeted by Donna Warnock of the Clamshell Alliance, who handed him two dead fish and said, "People are going to be outraged! You couldn't have found a better way to destroy the estuaries of New Hampshire than in the cooling system that you approved."

On Saturday, the Associated Press carried a story quoting Costle as saying "that the White House made no attempt to influence [my] decision. The President and the White House did not review the decision or judge the decision. It was essentially my decision."

That statement was tantamount to William Westmoreland arguing that he attacked Southeast Asia on his own.

As the news of Costle's decision went over the wires, a hundred nuclear opponents marched to the Seabrook access road, dumped five buckets of clamshells, and tacked a "Will Not Be Built" placard below the official sign reading "Seabrook Station." In a statement endorsed by twenty antinuclear groups from around the U.S., Kate Walker described the Seabrook decision as "a Declaration of War against the natural environment and those dedicated to protecting it. Jimmy Carter may be signaling with this decision that he will make himself the Lyndon Johnson of the 1970s."

Among other things, Walker ran through a long list of demonstrations and occupations scheduled for nuclear plants in Europe and the U.S., including an occupation at the Diablo Canyon nuke in California on August 7, and decentralized actions throughout New England August 6–9.

"Carter's decision makes another occupation of Seabrook almost inevitable," added Rennie Cushing, who has been arrested at the site three times.

Indeed, Carter's decision has made a whole lot more almost inevitable. The governmental approval at Seabrook means there is no intervenor recourse on earthquake, thermal discharge, impossible evacuation, or local opposition grounds. The Carter administration obviously feels the plants are needed, and that a design such as Seabrook's is basically acceptable. So if there is a "last resort" available to those who would stop nuclear proliferation, it is clearly not to be found in the nuclear policy as handed down from the newly elected President, a policy substantially—but not surprisingly—in conflict with that on which he ran.

Or, to put it another way: Here we go again.

☆☆☆

Soon after the Carter Administration approved the Seabrook cooling system, supporters of the project staged the nation's biggest pronuclear rally. Organized by an ad hoc coalition called New Hampshire Voice of Energy, the demonstration featured a march through the streets of Manchester and a mass gathering at the John F. Kennedy Arena. More than three thousand people showed up from New York, New Jersey, and New England, most of them construction and utility workers who donned hard hats, unfurled union banners and picket signs and marched through the streets chanting "Nukes! Nukes! Nukes!" Many bought T-shirts reading "Nuclear Energy: Safer than Sex," and bumper stickers bearing the slogan "Nuclear Power: The Pollution Solution."

The crowd at the coliseum heard speaker after speaker denounce the nuclear opposition as opponents of economic expansion. "This really is a death struggle against the no-growth advocates," said Edward J. King, a Massachusetts business leader who would, in 1978, become governor of his state.

Keynote speaker was MIT Professor Norman Rasmussen, author of the NRC's study on nuclear safeguards and a director of Northeast Utilities. The people "who call themselves environmentalists," he told the cheering crowd, were "irrational and illogical." Their purpose was to "threaten the jobs and livelihood of the region."

An unannounced speaker at the rally, and the man who drew the biggest ovation, was Meldrim Thomson. "[You] good

Americans" who filled the hall, he said, "came together, obeyed the law, and made your point. You're much better than what I saw May first. By comparison, you're beautiful."

Nuclear opponents charged that the rally had been staged and paid for by utility and industry supporters. According to the Associated Press, New York and Rhode Island power companies had paid for buses and food to encourage employees to attend. The PSNH itself had subsidized the rally with a grant of $1500. Company representative Norm Cullerot assured *The New York Times* that the money had come out of stockholder rather than ratepayer funds. "It's good to see the silent majority being heard," he added.

The demonstration underscored ongoing support for the Seabrook project, and further reminded antinuclear organizers of the urgency of building ties with the union movement (Chapter Nine).

It also contributed to a rethinking of strategies. Though enormous antinuclear momentum had been generated around the health and safety hazards of atomic power, economic issues now began to take the limelight. Throughout the United States utilities committed to nuclear power were raising rates to pay for their expansion programs. The hikes prompted widespread ratepayer anger that led to spontaneous uprisings against the utilities. Within months the "'rate revolt'" had crippled atomic construction programs throughout the Northeast:

People against Power
(*The Progressive*, April, 1978)

IN OCTOBER, 1977, more than two hundred fifty angry residents of Hartford, Connecticut, swarmed into a public utilities hearing, demanded a halt to high electric bills, shouted their opposition to a proposed new $90 million rate increase, and finally forced the commissioners to call in the police.

In December, New Hampshire's harassed public utility commissioners fled through a side door when Donna DiAntonio, a smiling young organizer, walked unannounced into a hearing loaded down with eight thousand "Vote No" postcards collected from around the Granite State to protest a proposed 17 percent rate increase.

Soon thereafter, New Jersey's utility commissioners suffered through a marathon thirteen-hour session at Vineland while four hundred angry ratepayers cheered speakers from consumer, labor, senior citizen, antinuclear, and community groups who attacked virtually every facet of the state's utility rate structure.

Such scenes have become common throughout the country. Soaring electric bills have brought on the beginnings of a grass-roots campaign that shows signs of shaking the utility industry to its core. Consumer resistance has already had a marked impact on general energy policy—and nuclear power construction—in the Northeast.

Electric utilities, which once took periodic and perpetual rate increases virtually for granted, now are confronted by an angry public determined to stop them. Regulatory commissions, which have acted for years as rubber stamps, now must cope with consumer resistance that has become increasingly well organized, and that is beginning to escalate into outright refusal to pay bills—a tactic with ominous but as yet untested implications for the future of private investor-owned utilities.

The problem is that without constantly raising rate revenues, utilities cannot attract new investors to massive new generating facilities. And without such constant expansion, energy policy in this country could undergo a radical transformation.

"We were sitting around the kitchen complaining about our electric bills," say Michael DiBernardo of Mantua, New Jersey. "We found the fuel surcharge was half the bill. So we looked into it and learned that the company was thowing all sorts of charges into it that shouldn't have been there. We agree that when their costs go up, the utilities should be compensated. But when they use the clause as a conduit for other things, that's wrong, and we decided to do something about it."

DiBernardo is president of Utility Users for Rate Reform (UURR), which emerged from his kitchen a year ago to attract some 44,000 dues-paying members (at $1 a year) with chapters in seventeen of New Jersey's twenty-one counties. "I never dreamed it would mushroom like it has," says DiBernardo. "This thing has really caught on."

UURR has now distributed more than 50,000 form letters used by ratepayers who want to withhold a part of their bill in protest.

DiBernardo say the utilities won't tell how many customers are withholding, but he is certain the number is in the thousands. Nor is it happening just in New Jersey. Consolidated Edison of New York has consistently complained about thousands of customers who routinely refuse to pay their bills, and withholding campaigns have recently been launched in West Germany, France, and Italy.

The tactic can be strictly legal. Under utility regulations common throughout the United States, a ratepayer may withhold all or part of the bill pending an appeal before the company and the state utility commission. Depending on the state of the bureaucracy and the number of people withholding, it can take days, weeks, or months for the company to turn off the juice.

There is no way as yet to gauge how badly a utility can be hurt, but New Jersey offers some indicators: "They keep sending notices, intimidating letters, and so forth, even though people are perfectly within their rights," says DiBernardo. "They even sent representatives—we call them 'Avon men'—to try and personally intimidate people into paying their bills. They wouldn't be doing this unless they were really hurting." According to UURR's Gina Secrest, the battle has escalated to the point where small municipal governments are refusing to pay their bills, and in retaliation the utilities are threatening to refuse payment for transmission-line passage rights.

The prime consumer target has been New Jersey's infamous fuel adjustment clause, which the state's utilities have used to inflate bills by including costs of transportation, administration, and a wide range of subsidiary charges, though the clause was meant only to provide for inflation fuel costs. Charges have been further inflated by the common utility practice of shuffling fuel through their own subsidiaries to multiply the final price.

Because these adjustment charges have been levied at the companies' discretion—outside utilities commission control—the bills to consumers skyrocketed. "I joined up when my bill hit $150 a month," says Secrest. "Our electric and gas are more than our mortgage payments, and half of it is in the adjustment clause. It's ridiculous. Why should we pay it?"

With UURR and other consumer groups in the lead, a bill requiring public hearings on adjustment charges has become law in New Jersey. A strong move is also afoot to force the election—rather than gubernatorial appointment—of public utilities commissioners.

The movement is not likely to stop there. "This is an issue that attracts an incredibly broad cross-section of people," says Susan Blake of the Long Island Safe Energy Coalition. "We're getting calls from retired people, working people, fixed-income people, middle-class people who have never been politically active. Nobody wants to pay the higher bills, and the fact is, a lot of people simply can't."

Blake's organization has been primarily concerned with stopping construction of a nuclear-power reactor at Shoreham. But in the spring of 1977, the Long Island Lighting Company (LILCO) demanded a 15 percent rate increase to build the plant. Antinuclear activists quickly joined a broad coalition aimed at a withholding campaign.

Organized withholding campaigns often ask that consumers put withheld funds in individual or group escrow accounts, but the Long Island campaign will simply inform people of the procedures involved in not paying. "We want to emphasize that this is more than a protest," says Blake. "People really can't afford to pay these bills."

Antinuclear groups have also joined with the UURR, for a number of New Jersey utilities are also demanding rate increases to build reactors. Particularly controversial are two floating reactors proposed (and recently postponed) for siting off the coast near Atlantic City. "I've learned a bit about nukes since we got into this," says DiBernardo. "And now I'm totally against them. We don't even know if they'll float, let alone operate, and the consumers shouldn't be made to pay the costs. A lot of them can't comprehend the dangers, but if you show them where the utilities are building nukes for the sake of boosting their rate of return, then you can show people where it's hurting them in their pocketbook, and this is the way it should be attacked from now on."

Part of the attack inevitably focuses on rate-base provisions. By law, utilities base their profits on how much money they invest in generating facilities. The more they spend, the more they can earn—and the more, ultimately, consumers must pay.

Another bone of contention is "phantom taxes," a practice by which the utilities charge consumers for taxes and then keep the money instead of paying it to the government. Some $2 billion was charged the rate-paying public in 1977 in the name of taxes, but then remained in utility coffers. Much of the alleged government tax was based on new construction costs, and at least some of the "taxes" that were collected—and then withheld—were used to promote still more construction.

But perhaps the most unpopular aspect of the utility rate game is Construction Work in Progress (CWIP), which has become an especially hot issue in New Hampshire. In the summer of 1977 the PSNH asked for a $32 million rate increase—about 21 percent—of which about half was CWIP for Seabrook. A bill outlawing CWIP easily passed the New Hampshire House but lost in the Senate, which is dominated by the state's pronuclear Governor Meldrim Thomson, who had promised to veto the bill if it passed.

The atmosphere surrounding the Seabrook project is tense: More than 1,600 protesters have been arrested at the site, and another mass civil disobedience action is planned for June 24.

Meanwhile, the company has made it abundantly clear that without CWIP and a long string of rate increases, the project will be in serious jeopardy. Company officials have admitted that building the plant may require rate increases of 5 to 8 percent a year until it is finished. "It's pretty amazing to hear them talk about rates going up 50 percent just to build a plant they say will lower rates," says Jeff Brummer, co-ordinator of the New Hampshire "Vote No on the Rate Hike" campaign. "We're getting all sorts of people calling in mad as hell who would never occupy at the site, but who sure aren't happy about their bills going up."

The "Vote No" organizers have distributed some 25,000 forms for protesting the increase. The sheets include preaddressed post cards—one for the governor, one for the Public Utilities Commission, and one for the campaign office files—demanding that rates not be raised. So far, Brummer estimates, at least 10,000 responses have been tallied. The campaign hopes to have chapters in all 270 New Hampshire towns, and has called for a state referendum on the rate increase which, says Brummer, "we would definitely win."

The campaign has made the plant's supporters edgy. Antinuclear groups, complained PSNH's William Tallman, are "preying on the perfectly natural resistance of consumers to price increases." William Loeb's reactionary *Manchester Union-Leader* added that "the professed concern of the antiprogress kooks for the people's pocketbooks is an obvious sham."

Though such rhetoric is familiar to readers of the *Union-Leader*, it has yet to help the PSNH's precarious finances. On December 3, 1977, the company posted a bond allowing it to raise rates pending PUC approval. If the PUC turns down the application (not likely) the PSNH could be forced into a devastating refund. But a spontaneous rate-withholding drive has already begun around the state, and the "Vote No" campaign will soon convert itself to a formal "don't pay" project. "The point is to hurt them, to bang into their cash flow," says Brummer. "If we can cut their quarterly dividend in half—by $2 million—I think we'll give the stockholders good reason for concern, and make any investors think at least twice about buying into Seabrook." All of which has prompted the PSNH's Tallman to threaten that if the flow of cash or rate increases is disturbed, "someone else" will have to build the Seabrook plant.

In nearby Connecticut, those angry protesters at the Hartford hearing were objecting to a $90 million rate increase demanded by Northeast Utilities (NU) for a massive construction program scheduled to include a 12 percent share in Seabrook, 75 percent of twin nuclear reactors proposed for Montague, Massachusetts, and a controlling interest in the Millstone III reactor being built at Waterford, Connecticut. Without all that power, said NU, the region could expect brownouts, blackouts, and worse. And without the $90 million, NU could not attract investors to pay for the construction.

Two of NU's three subsidiary companies rank first and third in the nation in excess-generating capacity. Some estimates put the company's real capacity at 70 percent beyond peak. Yet the company was demanding money for three nuclear projects. Public Utilities Control Authority (PUCA) hearings around the state played to packed, tense halls, as a long string of witnesses testified to the fat in NU's capacity, and to the soaring costs of building and operating nuclear power plants. According to Alden Meyer of the Connecticut Citizens' Action Group (CCAG), "the commissioners were finally forced to realize that nukes are getting priced out of the market, and may no longer even be competitive with oil."

They also learned that NU's predictions for growth in demand were optimistic, at best. The company had originally based its construction program on a demand growth rate of about 7 percent. But last year the company revised its estimates down to 4.7 percent, and it is now talking in terms of 3.5 percent. Consumer groups contend that even this is a bloated estimate.

In light of these figures, and in recognition of a political ground swell, the PUCA slashed NU's $90 million request to $35 million. It ordered the company to sell its share in Seabrook, forget about building the reactors at Montague, and complete construction at Millstone III as soon as possible. "We really can't afford to bankrupt a whole generation of ratepayers by a construction program which is building an edifice that is not required and may never be required," said PUCA member David Harrington.

Though the PUCA decision was well received around the state, some groups are now organizing a withholding protest around the $35 million increase that was granted. And NU, too, was unhappy about the decision: Instead of moving ahead with Millstone III, the company laid off the 1,200 workers there. It also put some 400 line and maintenance employees on furlough, threatened a cutback in service during storms and outages, and advised the public that it could now afford to install "only low-quality wire."

"They're obviously playing hardball with their ratepayers," says Commissioner Harrington. "They better believe the authority will play hardball with them. The only cut I have not heard about is in the salaries of the top executive officers."

NU's layoffs, added a bitter Governor Ella Grasso, had "apparently been designed to discredit the work of the PUCA." The issue has now escalated to questions of "corporate blackmail" and who is really in charge, the company or the government.

A bit to the north, Boston Edison has also postponed plans for its Pilgrim II reactors at Plymouth, pending a decision on its request for a $69 million rate increase. The company can't really prove a need for the facility and will face a withholding campaign—already in its early stages—if it does get the increase. New England Power, also having trouble, has postponed its plans for twin reactors at Charlestown, Rhode Island, leaving Seabrook and Millstone III the only nuclear plants under construction in a region once slated to be the showpiece of the new technology.

"Everybody is learning that the main reason rates are going up is because of expensive construction which just isn't necessary," says Rick Morgan of the Washington-based Environmental Action Fund.

In January, the EAF published a major report indicating that seventy-eight of the nation's top utilities had generating capacity significantly in excess of the 15 to 30 percent reserve required by federal law—a reserve Morgan terms "too big" anyway.

EAF, like a growing number of grass-roots consumer groups, is pushing for public power as a solution to the expansion problem. "Every time utilities go for rate increases, they have to get a higher return than the last utility in order to attract investors," says Morgan. "And they *love* racing against each other, because they can use the capital shortage as an excuse for higher profits. And as long as they can put construction in the rate base, or duck their taxes, or get CWIP, they'll be caught up in this growth thing whether it's good for society or not."

But the continued expansion inevitably brings higher electric rates, and it seems clear that public tolerance is rapidly approaching the point of no return. "Name me a state and I can name you people who are fighting the rate structure," says Morgan. "This is the ideal issue to bring together labor unions and consumer groups and environmentalists."

"There wouldn't be rate increases if it weren't for that unnecessary construction," adds New Hampshire's Jeff Brummer. "Jimmy Carter gave us the moral equivalent of war, and we've all seen what that's got us. Now let's try the economic equivalent."

☆☆☆

Underlying much of the burgeoning antinuclear campaign was an interest not only in safe energy, but in new ways of bringing about social change. Nonviolence became a major issue of discussion alongside nuclear power. For many, it was clear that given the civil liberties side-effects of an atomic society, violence and nuclear energy could never mix. It also seemed evident that peaceful resistance could offer organizers a new foundation on which to build a lasting, mass movement for social change. In either case, the tactic seemed to offer a "soft path" toward a new energy future:

The Power of the People:
Active Nonviolence in the United States
(*New Age*, September, 1977)

NONVIOLENT CIVIL DISOBEDIENCE is quintessentially human and utterly revolutionary. It is a tactic meant to combine the best of the political world with the highest of the spiritual.

It dates back to Christ and undoubtedly earlier. In white America it goes back to the 1600s, at least to Roger Williams, the laborer and minister who was booted out of Massachusetts for (among other things) insisting that land could actually belong to Indians. Also exported were Anne Hutchinson and other Quaker Friends who insisted that conscience was higher than the law, and then steadfastly refused to take up arms against the state.

As early as the 1640s, Dutch Mennonites shunned weapons of any kind, and in 1658 Richard Keene, a Maryland Quaker, was cited by his local draft board for refusing to be trained as a soldier. "You dog," said the local sheriff. "I could find it in my heart to split your brains."

Actually, worse was done to other heretics whose presence was so obnoxious to the Bay Colony that anyone who brought in a Quaker was subject to a fine of forty shillings per hour.

What made their presence so fearsome was moral rather than physical force. For although they shunned violent conflict, the Friends and other activist sects did not shun politics. Far from it—they attacked injustice wherever they saw it, which was almost everywhere.

But they went on the offensive with a very amazing set of principles. "I will suffer," they said in essence, "until you, the oppressor, give way. I will deny myself of what you have until it is given communally, as it should be. I will let you do whatever you want, but you'll have to take responsibility for my body to do it."

In essence, the assumption is counter to all the basic assumptions of traditional politics. It aims straight for the core of humanity in the opposition, and it doesn't let go until the best comes to the surface.

Realpolitik, the world of Metternich and Kissinger, assumes the worst in people. You build your power bases, bristle with guns and knives, and assume your opponents are bullies who'll only let go when their lowest instincts are satisfied.

Peaceful resistance works in exactly the opposite way. In a sense, it's unfortunate that the tactic has been so frequently labeled "non-violent," for the two negatives don't quite tell the whole story. "Love was the first motion," explained Quaker John Wollman, "and thence arose a concern to spend some time with the Indians, that I might feel and understand their life and the spirit they live in."

"We cannot acknowledge allegiance to any human government," added William Lloyd Garrison two centuries later. "Our country is the world, our countrymen are all mankind. We register our testimony, not only against all war—whether offensive of defensive, but all preparations for war, against every naval ship, every arsenal, every fortification, against the militia system and standing army; against military chieftans and soldiers; against all monuments commemorative of victory over a foreign foe, all trophies won in battle, all celebrations in honor of military or naval exploits; against all appropriations for the defense of a nation by force and arms on the part of any legislative body; against every edict of government requiring of its subjects military service. Hence, we deem it unlawful to bear arms or to hold a military office."

Furthermore, said Garrison, "we cannot sue any man at law to compel him by force to restore anything which he may have wrongfully taken from us or others; but if he seized our coat, we shall surrender up our cloak rather than subject him to punishment."

Noble sentiment indeed, but what distinguished Garrison and those around him in the abolitionist movement was their willingness to act on their beliefs, and act effectively. "Be assured," he said, "that until your cause is honored with lynch law, a coat of tar and feathers, brickbats and rotten eggs—no radical reform can take place."

Indeed, along with thousands of others, Garrison constantly put his body on the line to win peace and freedom for all. There were those who refused to fight the native Americans, and those who abstained from the wars against the British. There were those who fought for the freedom of slaves in the South but just as steadfastly refused to participate in the Civil War. There were the Grimké sisters, pioneers of the women's rights movement, and Eugene Debs, the Socialist labor pioneer, who recognized the common consciousness of all humanity. "When I see suffering about me," he said, "I myself suffer. And so when I put forth my efforts to relieve others, I am simply working for myself.''

In 1918, sitting before a judge who was about to give him ten years for opposing World War I, Debs expounded: "Years ago I recognized my kinship with all living beings, and I made up my mind that I was not one bit better than the meanest on earth. I said then, I say now, that while there is a lower class, I am in it; while there is a criminal element, I am of it; while there is a soul in prison, I am not a free man.''

Like Garrison and the Grimkés, Debs was more than eloquent—he was an immensely effective politician, able to translate words into power. "The practical person,'' explains Cesar Chavez, "has a better chance of dealing with nonviolence than people who tend to be dreamers or who are impractical. We're not nonviolent because we want to save our souls, we're nonviolent because we want to get some social justice for the workers.''

There's the trick. Passive resistance, if it is to have an effect on the world, must be framed in a political context. There is, indeed, plenty of room for the individual statement of conscience. A case can be made that it is ultimately an engine of social change. But to be translated powerfully to the material plane, the force has to be focused, *organized*. "If all you're interested in is going around being nonviolent and so concerned about saving yourself,'' says Chavez, "at some point the whole thing breaks down—you say to yourself, 'Well, let *them* be violent, as long as *I'm* nonviolent.' Or you begin to think it's okay to lose the battle as long as you remain nonviolent.

"The idea is that you have to *win* and be nonviolent. That's extremely important: What do the poor care about strange philosophies of nonviolence if it doesn't mean bread for them?''

In the teens and the thirties, Wobblies and CIO auto workers sat down in "their'' factories, stalemating the owners. In the 1890s and 1920s dissident laborers followed factory regulations to the letter of the law, destroying production schedules and winning demands. In the fifties, blacks refused to ride buses in some cases and refused to get off them in others. In the 1860s and the 1960s, men refused to cooperate with the draft; citizens rejected war taxes and blocked troop transports. In the seventies we find ourselves marching onto nuke sites, searching for ways to block rampant military insanity, and looking for solutions to many of the same conflicts confronted by the first Quakers.

The point is not to be sanctimonious or holy. It is rather to be consistent and at the same time powerful, peaceful, and yet reach the goal. ''At the center of nonviolence stands the principle of love,'' wrote Martin Luther King. ''The end is redemption and reconciliation. The aftermath of nonviolence is the creation of the beloved community, while the aftermath of violence is tragic bitterness. Along the way of life, someone must have sense enough and morality enough to cut off the chain of hate.''

That chain includes not only emotions, but economic inequality, racism, sexism, imperialism, pollution, disease, and all the other traditional political ills that have plagued human society.

''Christ showed us the way,'' wrote King, ''and Mahatma Gandhi showed us it could work.''

What remains for us is to do the work and see how far the tactic can take us.

5
Seabrook 1978:
The Movement Hits
the Mainstream

☆☆☆

Resistance Gets Set for Spring
(*The Nation*, February 11, 1978)

AT 9:30 A.M. DECEMBER 18, 1977, an explosion rocked the Millstone
I nuclear reactor at Waterford, Connecticut. Hours later, while plant
officials were on the phone to the Nuclear Regulatory Commission, a
second explosion blew an eighty pound steel door 130 feet through the
air and into a nearby warehouse, contaminating a storeroom super-
visor.

The blasts added new bitterness to a confrontation that in recent
weeks has brought the nuclear industry in New England close to
paralysis, witnessed key trials of antinuclear activists across the
United States, and marked the announcement of spring and summer
dates for a series of critically important civil disobedience actions.
Indeed, these latest "incidents" at accident-prone Millstone could
hardly have come at a worse time for its owner, Northeast Utilities
(NU), or for the nuclear industry.

The Millstone I explosions could have contaminated construction
workers at the nearby Millstone III site. As it was, they were lucky

enough to have been laid off at the time. That, however, did little to mollify Governor Ella Grasso and other state officials who were infuriated by NU's failure to notify them promptly that the explosions had occurred. In the wake of earlier mishaps, Grasso had ordered NU to inform her office within an hour of any incident at Millstone. As it was, she wasn't told of the morning's blast and radioactive "puff" until late in the afternoon.

Grasso's anger was echoed four days later in the north, when news broke through the Associated Press that a similar explosion at the Vermont Yankee plant at Vernon had gone entirely unreported. Occurring on December 10, the incident became public knowledge only when plant officials casually mentioned the incident to a member of the state public service board. "If the technical competency of Vermont Yankee was as bad as its public relations," said the state's pronuclear Governor Richard Snelling, "we'd all be in the dark."

And the nuclear industry's winter of bad news was hardly limited to these explosions. On the day of the Millstone blowout, Boston Edison leaked word that it was postponing for the third time its target date for opening its second reactor at Plymouth. Citing a slump in the demand for electricity and the prospects of rate reform, the announcement was accompanied by the news that the Nuclear Regulatory Commission (NRC) had denied Boston Edison a conditional construction permit, thus making it more difficult for the company to get the project built to a point where public resistance would probably not halt it.

A week earlier NU had formally acknowledged its second postponement of the Montague project, and revised its "on-line" target date for Millstone III to 1986. New England Power Company also joined the crowd by pushing back its target date for twin nukes planned at Charlestown, Rhode Island.

Further down the coast, the Public Service Company of New Jersey announced a three-year delay in its plans to float reactors in the ocean north of Atlantic City. In Louisiana, the utilities commission has denied nearly all of Louisiana Power and Light's $55 million rate-hike request, thus projecting a dubious future for the company's Waterford III reactor, now under construction. And in California, Governor Jerry Brown's energy commission slashed one of two reactors proposed at Sundesert, and made the other at best "iffy."

But though the pattern of postponement and uncertainty spreads across the country, the reverses have been heaviest in New England.

The region was once considered atomic power's prime territory. But with every proposal for a new plant in doubt, and with the one under construction in disarray, the region has become a proving ground for the antinuclear movement. And nowhere is that drive more strongly organized than at Seabrook, New Hampshire.

The site has already been the scene of three mass nonviolent occupations. In early November, the Clamshell Alliance, which organized the occupations, set plans for a fourth—this one promising to be the biggest mass civil disobedience march in this country in many years.

Meeting at Putney, Vermont, the November congress was attended by some three hundred representatives from more than fifty Clamshell groups. The weekend meeting set June 24 as the date for the next direct-action, nonviolent attempt to halt construction. The Clamshell vowed also to restore the devastated 750-acre Seabrook site to at least a semblance of its natural state, and to install natural energy equipment as a symbol of the movement's commitment to renewable forms of power. All those wishing to occupy would undergo special training in nonviolent techniques. Reflecting a deepened commitment to neighborhood organizing, the alliance also specified that occupiers are to engage in door-to-door canvassing before coming to take the site.

The Clamshell decision to reoccupy was the more dramatic because sixty-four of its members who had occupied the site last April 30 were due to face trial at Rockingham County Superior Court the next day. The trials had been demanded by the alliance as part of the membership's wish to test in front of a jury the merits of civil disobedience in the context of the dangers of atomic energy.

The first case was called on November 8 with Carter Wentworth, twenty-six, an artist from nearby Kensington. He had agreed to be the focal point of the Clamshell's attempt to "put nuclear power on trial." The New Hampshire judiciary, however, was unwilling to go along.

Wentworth and his lawyer, Eric Blumenson of Boston, had intended to make their case on the New Hampshire "competing harms" statute. The law states that an individual is justified in breaking the law if he or she is attempting to prevent a damage greater than would ensue from the illegal act. Swerving illegally to avoid a pedestrian is a common example of the situation. However, when the statute was raised by Blumenson, Judge Wayne Mullavey ruled it out of order.

After swearing in the three-man, three-woman jury, Mullavey next barred a long string of witnesses who were to testify on the dangers of nuclear power, the politics of nuclear licensing and the history of civil disobedience. Among those not allowed to speak were Drs. Helen Caldicott and Rosalie Bertell, experts in the health effects of radiation; Professor Howard Zinn, who has made a study of civil disobedience, and Tony Roisman, a Washington attorney who has represented people opposing the Seabrook permits.

A series of character witnesses did testify, however, as did Clamshell members, who explained to the jury the aims and procedures of the Seabrook occupations, including a long analysis of the meaning of nonviolence and the commitment to stopping nuclear power. Finally, Wentworth took the stand. "I went to Seabrook to protect my life and my neighbors' lives," he said. "I was acting under the freedoms given to us in our Constitution and the Declaration of Independence."

Although Mullavey had denied the relevance of the competing harms law, Wentworth and the Clamshell witnesses continually referred to it. It thus came as no surprise when the jury, during deliberations, asked Mullavey for a copy of the statute. Instead, the judge sent in his own interpretation of it, which varied significantly from the original. In his charge to the jury, Mullavey also misinterpreted the criminal trespass statute itself, neglecting to emphasize that the defendant must "knowingly" commit the crime.

The jury brought a verdict of guilty. "Had we been able to call our witnesses, and had the jury been properly informed of the competing harms and criminal trepass laws, we're sure Carter would have been found innocent," said Robin Read, speaking for the alliance. Mullavey, however, had more up his sleeve. Ignoring prosecution requests for a fifteen-day sentence and a $100 fine, he pounced on the opportunity to slap Wentworth with four months in prison. It was the stiffest sentence for criminal trespass handed down by Rockingham County in at least twelve years. The occupation, Mullavey said, had been "mob action," producing "an explosive situation. This is one of the few cases since I've been on the bench in which sentencing may serve as a deterrent to future crimes of this type."

Trials held under similar conditions in the adjoining courtroom of Judge William Cann resulted in two- and three-month sentences for Carolyn Dupuy, a Hartford nun, and Roger Cole, a former local football hero. Court Dorsey, a street musician from DeKalb, Illinois, followed Wentworth in front of Judge Mullavey, conducting his own defense. He, too, received a guilty verdict and three-month sentence.

Mullavey's hostility and stiff sentences set the stage for the alliance's fourth occupation announcement. "Unjust sentencing like that won't scare people away," said Clamshell representative Cathy Wolff. "If anything, it'll bring more."

Shortly thereafter, twenty-two former occupiers were sent to various New Hampshire county jails for refusing to pay $100 fines resulting from their arrests. Most of them spent twenty days. One group managed to screen *The Last Resort*, a documentary film about the Seabrook occupations, in the Stafford County jail at Dover.

Although the Superior Court verdicts were hailed by plant supporters as a victory, it was at best a pyrrhic one. The court had presided over "nuclear power on trial" by refusing to hear evidence against nuclear power—in perfect accord, the Clamshell said, with the way the entire public debate has been conducted. In addition, the court could dispose of just four cases in more than a week of sittings. Some 1,500 Seabrook-related cases remain to be tried, and that is now a very hot issue in a county that has become both increasingly antinuclear and increasingly concerned over the cost of trying and jailing so many occupiers.

Meanwhile, parallel trials were going on throughout the United States with varied results. At Brattleboro, Vermont, eighteen members of the Vermont Yankee Decommissioning Alliance were fined $25 each for blocking the Vernon plant gates on October 5. In California, members of the Abalone Alliance were involved in a far more complex legal battle. Fifty-one had been arrested last August 7 while occupying at the Diablo Canyon plant in San Luis Obispo. Their initial sentences, handed down by a local judge, included a rare mandatory $500 fine. But in the midst of proceedings, the Abalones discovered that their ranks had been infiltrated by at least two police undercover agents, one of whom was prominent on the alliance legal committee. Richard Lee, who proved to be a local deputy sheriff, was "very active in our defense," says Raye Fleming, an Abalone founder and occupier. "It's hard for us to understand what could motivate a person to be that dishonest with himself."

Whatever motivated him, Lee's infiltration may have resulted in "leaks" from the Abalone defense committee—which, in turn, may result in the charges being dropped. The case is now being appealed on the ground, among others, that the infiltrations flagrantly violated attorney-client privilege and voided the Abalones' chances for a fair trial.

Further up the West Coast, members of the Trojan Decommissioning Alliance (TDA) engineered a trial that was everything the Clamshell trials were not. Eighty-one TDA members occupied at the Trojan plant, some 30 miles northwest of Portland, Oregon, on August 6, 1977. The day after Thanksgiving, 124 repeated the act. In a five-day trial that ended December 16, ninety-six TDA defendants were found innocent of trespass charges. At the outset, Columbia County District Judge James Mason allowed expert testimony into the case, with the reservation that it might later be stricken from consideration. Although Mason would have preferred to ban the testimony altogether, TDA's Norman Solomon said, "That was part of the deal. Unless the judge allowed our experts to testify, we were going to go for individual trials, which would have cost the county a fortune."

Thus Drs. Bertell and Ernest Sternglass lectured the jury on cancer rates and radiation levels surrounding nuclear power plants. Lon Topaz, a former director of the state Department of Energy, gave angry testimony against the Trojan reactor, calling it "an imminent danger" to human life that "should be immediately shut down and decommissioned."

The defense testimony received statewide media coverage during the trial. It did not, however, prevent Judge Mason from instructing the jury to consider neither the state's "choice of evils" law, nor the nuclear-related testimony in their deliberations. The jury deliberated for five hours and delivered a "not guilty" verdict based on a technicality stemming from the fact that the occupiers were arrested on railroad rather than utility property. A poll of jurors showed that, had they been permitted to recognize the choice-of-evils law, the not guilty verdict would have come "in five minutes."

Even at that the TDA's Margot Saegre hailed the verdict as "a great victory," and the Portland General Electric Company, Trojan's owner, seemed to agree. The decision, said public relations director Steven Loy, would probably "create some enthusiasm" for new occupations. According to TDA sources, another one is being planned.

Indeed, through the widely divergent conduct and outcomes of the various trials, one theme is clear—most or all of the occupiers seem intent on returning to their neighboring nuclear sites. Early in May, simultaneous civil disobedience actions are planned at the Barnwell-Aiken weapons waste-storage facility in South Carolina, and at the Rocky Flats weapons plutonium-processing plant in Colorado. Later in the month, the Mobilization for Survival, a national coalition, will sponsor a mass march on the United Nations to demand nuclear disarmament. By the time of the June Seabrook occupation, more than a score of other antinuclear actions are likely.

Thus, 1978 promises to be a bad year for the nuclear industry. Faced with legal challenges at all levels of the courts on topics ranging from insurance to fuel supply to plant effluents to construction costs and sagging electrical demand, the industry may be providing as many jobs for lawyers as for construction workers.

With explosions like those at Millstone and Vernon becoming increasingly noticed, with rate hikes coming under sharp attack, with postponements of construction increasing, and with just seven new domestic orders logged in the last two years, the nuclear industry must indeed see 1978 as the year that could make or break it.

But despite its hard time, the industry is not likely to surrender. An assault on state siting councils and utilities commissions, a drive for licensing easements, and massive lobbying for huge federal tax credits and subsidies are the least we can expect from both reactor producers and utilities.

If the trials and new occupation plans of antinuclear activists are any indicator, 1978 could also witness a matured nationwide anti-nuclear campaign. Part of that maturity will come in recognizing that, while atomic energy may well be on the defensive, it is still far from open retreat.

☆☆☆

As the training for the fourth occupation proceeded, serious problems arose within the Clamshell Alliance. One group of Clams argued that since the site was now fenced, cutting holes through the chain-links might be necessary. If the purpose of the occupation was to stop construction, they said, then gaining access to the site was an obvious priority.

Much of the rest of the Alliance disagreed. Some felt that destruction of property should itself be considered a violent act. Others argued simply that in a mass situation, the fence-cutting might touch off an unwanted confrontation with police or construction workers—a potential powderkeg with unpredictable consequences. Given the conservative, volatile political climate in New Hampshire, such a confrontation might go a long way toward undercutting the Alliance's hard-won mass base.

Debate dragged on interminably, largely because of the Clam's unique decision-making process.

The Clamshell Alliance had no officers, no elections, and no titles beyond "staff" and "spokesperson," which for many Clams was already going too far. Workaday tasks were handled by finance, resource, media, legal, alternative energy, outreach, action, and other committees composed of volunteers from around the region. Personnel holding the poorly paid staff positions in Portsmouth and Boston turned over rapidly as nervous systems burned out under the strain.

Though it was nominally a regional coalition, the Clamshell Alliance really only existed as a unit for actions at the Seabrook site. Otherwise it remained a loose collection of local groups, each focused on the communities from whence they came. In fact, the Clamshell could disappear—as it usually did after major actions—and the same organizing would continue in town after town.

When centralized decisions were necessary, they were made at irregular region-wide congresses, or at more frequent coordinating committee meetings, which were peopled by representatives from the various local areas. The co-ordinating committee was essentially forbidden to make important decisions without either sending back to local groups for approval or calling a congress.

Clamshell meetings at any level could test the commitment of even the strongest Quaker soul. Although the very early Alliance organizing meetings operated otherwise, the Clamshell soon gravitated to an understanding whereby major decisions required "consensus," which was understood to mean absolute unanimity. A single dissenting voter could block an entire organizational decision. Since Clam meetings were always open, this meant—literally—that a total stranger could walk in off the street and stop the group in its tracks.

The process reflected a profound communal faith that had proved a great boon to the Alliance as it was forming. The necessity of gaining unanimity on major decisions forced the welding of a diverse but relatively small and friendly group of people into a unified organization. Every Clam was more than free to speak her or his mind, and no one could feel slighted by any decision made. Inevitably, when confronted with an important choice, discussion would carry on for hours until all points of contention were satisfied. A facilitator—meeting coordinator—would then ask for objections. The group would hold its breath, watching for dissenting hands.

When there were none, there was an involuntary cheer, and the decision was made. Thus the Clamshell moved forward by silent consent rather than noisy approval. Somehow, that created a tremendous cohesion and strength.

But the success of the process depended on relatively small meetings and a fair like-mindedness among the membership—or at least a consistent willingness to compromise. Consensus could always work in an affinity group or a small local chapter. But regionally the process might be expected to break down at that precise moment when those who dreamed of sparking a diverse mass movement actually succeeded, thus flooding their meetings and making absolute unanimity a dream of the past.

June 24, 1978, would mark just such a watershed.

The Rath Proposal

While the debate over fence-cutting dragged on, the Clamshell got a surprise from the government. It came in the form of an offer from Thomas Rath, who had replaced David Souter as New Hampshire's attorney general. (Souter was given a judgeship by Meldrim Thomson soon after the 1977 Clamshell occupation.)

Rath's offer seemed to come straight from Washington. It was simple enough. The Clamshell could use a piece of PSNH property on which to stage a legal rally. In return, the Alliance had to agree to leave the premises after a given time. Thus the "occupation" would take place on the Seabrook site; but it would make no attempt to stop construction of the plant or to breach the law.

The offer hit the Clamshell like a thunderbolt. The stated purpose of the occupation had been to stop the plant. The Rath proposal was clearly an attempt at co-optation.

But the compromise had clear benefits. It would allow nuclear opponents more time to organize around the rate-hike issue, which was by now quite obviously a very serious threat to the PSNH and the Seabrook project. It would undo much of the polarization on which Meldrim Thomson had thrived so long. And it would create an opportunity for outreach to the more conservative local opponents who would never break the law, but who might join with the Clamshell in a legal rally, especially if it were actually held on the Seabrook site.

Polarization within the alliance soared sky-high. Demands for rejection and counterdemands for acceptance combined with dozens of complex compromise solutions to destroy nervous systems throughout the region. For long, draining weeks the Rath proposal totally consumed the Clamshell's time and energy. All organizing stood still as the futile struggle for consensus carried on.

Finally it boiled down to this: since consensus had already been registered in favor of the occupation, unanimity would also be required to call it off. And that was out of the question.

So at a raucous twelve-hour meeting of the co-ordinating committee three weeks before the planned occupation, the Alliance "accepted" the Rath proposal—on condition that the PSNH prove the plant was safe and necessary, and that construction stop until that proof was made. Shutdown was the bottom line. The Clam would refrain from physically trying to stop construction only if the PSNH would agree to stop it for them.

The "acceptance" was universally greeted as a rejection, which it was. Another confrontation was in the wind. Rath said he was "deeply sorry" and began preparing for mass arrests. The *Union-Leader* crowed that Thomson and the PSNH had already won a "great victory" and that no matter what happened June 24, the Alliance had irrevocably alienated the people of New Hampshire. The Clams, added an elated Meldrim Thomson, were a "filthy, foul, and un-American minority," a "gurgling, spurting bunch of nonproductive individuals."

More serious was the rhetoric of Carlton Eldredge. The Rockingham County prosecutor had long claimed to be an opponent of the Seabrook project. "Until they can prove it safe," he said, the plant shouldn't be built.

But mass arrests were another matter. The Clamshell had cost the county thousands of dollars and its prosecutor countless hours of sleep. This time there would be no arrests, no armories: there would be dogs, tear gas, fire hoses, and "lethal force" to greet the occupation.

Out-of-staters greeted Eldredge's pronouncements with scorn. Locals were not so sure. Between Thomson and Eldredge, rationality on the part of the state could not be guaranteed. Colonel Paul Doyon had retired and been replaced as head of the state police with a Thomson appointee. The Governor himself had recently visited South Africa and lauded its racial policies as being among the most enlightened in the world. An aggressive politician with national ambitions, Thomson seemed eager and excited over the prospect of a battle in his own back yard being televised world-wide.

The people who lived in that back yard were less pleased. The Clamshell's rejection of the Rath proposal deeply stung some long-time supporters, who insisted it should have been accepted. "We have to live here," said Diane Garand, a Seabrooker who had agreed to allow the Clamshell to use her chicken farm as a base of operations. "Going ahead with the occupation is just playing into Thomson's hands."

Thomson's hand had already been evident throughout the town. Tax assessments for local nuclear opponents had jumped far higher than those who supported the plant. Known Clamshell sympathizers found themselves with zoning hassles, threatening phone calls, tax problems, and an escalating atmosphere of intimidation and potential violence.

They also saw little value in another wave of mass arrests. The last one had been effective, but it had cost the state a fortune. The Rath proposal seemed reasonable. It was time, said the locals, to "try something else."

As June 24 approached, local supporters pulled back. The use of land on which the Alliance had counted for staging areas was denied. The local opponents of the Seabrook plant wanted a mass get-together where "Closet Clams," who opposed the project but were unwilling to associate themselves with mass lawbreaking, could come out into the open. "I have neighbors, I see them in the store," Garand told a crucial late-night session prior to the action. "I look at them and wonder, 'Are they against the plant?' Well, now is the time I'd like to find out."

By Saturday, June 10, nearly all the local opponents had withdrawn their support of the occupation. They wanted, they said, a legal rally where citizens of the seacoast could get together and count heads.

And they got it. At a smaller-than-usual Clamshell co-ordinating committee, which had come together expecting merely to finalize logistics for the mass confrontation, the decision was made to go legal. Consensus was dubious. One representative quit. Another didn't quite know what to say.

But they all faced one simple fact: just four acres of New Hampshire land (the Santasuccis') was still available from which to stage the occupation. The three to five thousand trained occupiers on which the Alliance knew it could count might have been able to move in from Massachusetts and Maine. But that was contrary to the founding principle of the Alliance, that, as at Wyhl, Germany, actions must spring from the local community. No land, no local support—no occupation. To carry on was to jeopardize a political base that it had taken a decade to build, in exchange for an action whose success was not assured and whose payback was not clear.

With not a little trepidation, the co-ordinating committee took the plunge, and sent a delegation to Rath. "The Clamshell has decided to hold a completely legal action and to not transgress the fenced-in construction area," it announced. "Anyone who does so is not a member of the Clamshell Alliance."

Unfortunately, the decision was not sent back to local groups for approval, as was the usual Alliance procedure. The committee had deemed that time was too short, and that the action had already been irreversibly vetoed by the seacoast supporters.

Like the decision to stage an alternative energy fair in October, 1976, the June 24 reversal evoked a storm of internal ill-will. Charges of "elitism" and "subversion of the process" alternated with strong hints at police agentry and outright personal attack. There was doubt, frustration, and anger, not a little of it the product of nervous systems thoroughly overtaxed by weeks of exhausting debate and years of intense organizing.

Thomson's Loebotomy

But the divisions within the Clamshell seemed to pale by comparison with what was going on at the Statehouse. In the days following the Alliance's acceptance of a legal rally, it became apparent that Thomson and the PSNH had gone along with the Rath proposal on the safe assumption that the Clamshell would turn it down. Now that the Alliance had accepted it, rancor reigned.

Throughout the week following the first Clamshell response to the Rath proposal—the "rejection"—Loeb had editorialized that the offer of a rally site on PSNH property should now be formally withdrawn. To keep the offer open, said the *Union-Leader*, was to court disaster.

When the Alliance then did accept, the *Union-Leader* turned apoplectic. "We hate to say 'We told you so,' but we did," said the paper. Because the state had not withdrawn the offer, "the Clams have achieved a propaganda victory of sorts by out-maneuvering both the state and the Public Service Company. It's not that the Clams are overly bright. It's just that when it comes to understanding their perverse mentality, state and company officials are just plain stupid."

As for Meldrim Thomson, his mentor's organ labeled him "a liar," and lampooned him as a carnival barker, leading scruffy batallions of nuclear opponents onto the treasured site.

Thomson himself spent two weeks sniping at the details of the state agreement with the Alliance. Disputes over parking, public access, and porta-johns became poor substitutes for a national emergency. The Governor had been upstaged.

In the seacoast, the air lifted like magic. And the weather held. June 24 dawned bright, sunny, and hot. On a peaceful, media-saturated Saturday, six thousand trained Clamshell oc-cupiers, their numbers swelled by a last-minute surge of sup-port, moved onto the site. Eighteen acres of bog, dump, and woodlot soon erupted with tents, meeting centers, roadways, a "Bridge Over Troubled Waters," a first-aid tent, a makeshift stage, food stands, windmills, literature tables, T-shirt conces-sions, a geodesic dome, a solar oven, and various displays on conservation, recycling, and the paraphernalia of an organic twenty-first century.

On Sunday the Closet Clams stepped in. More than twelve thousand people—many of them conservative seacoast nuclear opponents who had never been to a rally before—swarmed onto the site. It was the largest crowd in the history of the American antinuclear movement to that time. It seemed to prove, once and for all, that the nuclear opposition could transcend its activist hard core and attract a solid mainstream audience. Alliance organizers were relieved and overwhelmed. Once again, the Clamshell had jumped through the eye of the needle:

We Did It Again!
(*Clamshell Alliance News*, July, 1978)

Seabrook

ON THE HOT, SUNNY SUMMER SOLSTICE WEEKEND OF 1978, thousands, thousands, and more thousands of Clams, new-found Clams, and "Closet Clams" amassed at the site of the Seabrook nuke.

Within a week, construction at the site was ordered to a halt.

Numbering some twenty thousand strong, the Clam occupation comprised the biggest antireactor gathering in the history of North America.

It signified—for both the Clamshell Alliance and the antinuke movement as a whole—yet another crucial quantum leap into the Solar Age.

Surrounded by Acres of Clams

The occupation/rally took place on an eighteen acre portion of the Seabrook site, within clear view of the giant cranes, cement factory, and support buildings that mark the spot of New England's largest construction project, the 2300-megawatt proposed twin reactor complex now "guesstimated" by its builder, the Public Service Company of New Hampshire, at $2.3 billion.

Unlike three previous Clamshell occupations, in which more than 1600 people have been arrested, this one was legal. Through complex (and controversial) arrangements negotiated with the state of New Hampshire, the Alliance took possession of eighteen acres of land at the Seabrook site on Friday evening, June 23, and left the afternoon of Monday, June 26. There were no arrests.

For 6000 trained, arm-banded Clams (6130 by official Associated Press count), the weekend occupation was filled with uplifting song and stirring marches, hot debate and hours of meetings, heated disagreement, and good prospects for reconciliation, love, friendship, unity, and, above all, renewed dedication for stopping the Seabrook and all other nukes.

For twelve to fourteen thousand members of the general public—many of them the "Closet Clams" who have supported the antinuke cause but not occupied the Seabrook site—the rally was a chance to see a working windmill, a geodesic dome, a wide array of solar collectors, a tremendous assortment of literature on the coming age of renewable energy, to hear superb speeches from energy experts and political activists ranging all across the national political spectrum, to get the low-down on the local situation from Seabrook town officials and Seacoast spokespeople, to witness a major breakthrough in the growing labor-solar alliance, to hear the music of Pete Seeger, Jackson Browne, John Hall, Arlo Guthrie, and others, to be present as the first indigenous power was generated at the Seabrook site—by the Clam windmill, Seabrook Unit 1—and to join in a weekend of peace on top of the nation's most hotly contested nuclear site.

Above all, the thousands who gathered at Seabrook gave proof of the fantastic growth in size and maturity of the antinuclear movement, leaving little doubt around New England that the Seabrook nuke can, in fact, be stopped.

From the eighteen New Hampshirites who occupied Seabrook on August 1, 1976, the force of nuclear opponents gaining access to the Seabrook site had multiplied a thousandfold in less than two years.

Not Much in Manchester

Meanwhile, across the state at Manchester, only a few hundred nuke supporters rallied to hear Gov. Meldrim Thomson make his pitch for radioactive energy. It was an embarrassing turnout by any standard, significant far beyond its paltry numbers. A year ago, more than three thousand had marched through Manchester in support of the plant. By 1978 their numbers had dropped tenfold. The nuclear industry has long claimed it has massive public support, especially from organized labor. But it was the Clamshell that really captured the public—and the support from organized labor.

Indeed, though Thomson later showed up at the Seabrook site to boast that the occupation had "failed to stop construction," within the week an order had come down that would do just that. By July 21, by order of the NRC, the wheels of destruction had ground to a halt at the Seabrook site, throwing the PSNH miles deeper into its already hopeless morass of political opposition, financial mismanagement, and self-generated regulatory confusion. For all its fancy lawyers and official arrogance, the company had failed to get a simple building permit.

Clams on the Move

As the action at the Seabrook site ended on Monday, June 26, more than 1700 Clams converged on the Hillsborough County Courthouse, where unique joint Environmental Protection Agency-Nuclear Regulatory Commission (EPA/NRC) hearings were underway.

During the vigil, one NRC Commissioner offhandedly remarked that the Clams might "Come visit us in Washington." And indeed, plans were already in effect to do just that.

Two days later more than 350 nuclear opponents set up camp outside the NRC offices in the Nation's Capitol.

By Friday, June 30—one week after the Clam preparation crews had begun moving onto the Seabrook site—the NRC issued an order suspending construction at Seabrook, seriously threatening the future of both the world's most controversial nuke, and the utility building it.

All Together Now

The "Seabrook Week" of June 23–30 brought together in an unparalleled way strength and solidarity from the women's movement, the black movement, the labor movement, the Native American movement, the environmental movement, all coming together with the basic struggle of the people of the seacoast, of New Hampshire and of New England to save our homes, natural ecology and lives.

There is still a long hard way to go in this struggle. But there was one clear and unmistakable message from the antinuclear activities of June 23–30, one that even the atomic industry will be hard-pressed to miss: THE SEABROOK NUKE WILL NOT BE BUILT!!

☆☆☆

...And Again
(*Clamshell Alliance News*, August, 1978)

Washington

THE NUCLEAR REGULATORY COMMISSION (NRC) got a small but potent taste of what the direct actions at Seabrook and around the country have been like, and the results for the antinuclear movement were pleasing to say the least.

On June 28, on the heels of the mass actions at Seabrook and Manchester, more than 350 nuclear opponents brought their case to Washington.

The action was organized by the Seabrook Natural Guard, an ad hoc coalition of members of the Clamshell and other antinuke alliances from around the country.

On Wednesday morning, the 28th, the "Guard" gathered on the steps of the Capitol, where they heard speeches from U.S. Representatives Toby Moffett (D-Conn.), Leo J. Ryan (D-Calif.) and Richard Ottinger (D-N.Y.), all of whom wished the marchers good luck and the nukes a speedy demise.

The antinukers then set out through the wilds of Washington. Their first stop was the Department of Energy (DOE), domain of nuke czar James Schlesinger. After songs, chants, a speech from Bob Alvarez of the Environmental Policy Center, and the "leafleting" of some members of the DOE work force, the march proceeded to Lafayette Park, across from the White House. Gratefully seated in the green of the park, the gathering swelled to more than a thousand, as the sympathetic and the merely curious joined the group from area office buildings. They then heard speeches from Daniel Ellsberg, D.C. City Councilwoman Hilda Mason, environmental attorney Tony Roisman, and Diane Garand, Shirley Gustavson, Joan Bigler, and Lu Gunderson of Seabrook, Hampton Falls, Kensington and Newton, New Hampshire, all of whom had come to register their local complaints about the Seabrook nukes.

Meanwhile, that morning, staff officers at the NRC had continued negotiations for a series of direct meetings with the commissioners. There are five seats on the NRC, but only four are currently filled, by Chairperson Joseph Hendrie, and Commissioners Richard Kennedy, Victory Gilinsky, and Peter Bradford.

In the wake of the massive antinuke actions at Seabrook and Manchester, Washington was full of talk of the coming decision on Seabrook, and it all seemed to crystallize around the presence of the Natural Guard and their move toward the NRC.

For all the chaos at the nuclear sites it regulates, the NRC had never been the target of a mass occupation.

And now the timing could hardly have been more crucial. The Seabrook construction permit was once again in limbo. When the NRC originally allowed construction to start at Seabrook, it had been conditional upon approval of the plant's ludicrous cooling system by the Environmental Protection Agency (EPA).

But a federal court had disallowed that EPA approval. Would the NRC now lift the construction license?

In the early afternoon, the Natural Guard occupiers moved from Lafayette Park to the NRC offices on Northwest H Street to get an answer.

What they found was enough to make the bog and dump of the Seabrook rally site seem like paradise. The NRC offices were perched high above the street, in a building whose only "lawn" consisted of a few brave bushes sticking out proudly between the office facade and the noisy, dirty street.

With the concrete soaking up the blazing heat like a solar oven, the NRC occupiers pitched camp, carefully delineating the occupation area from the sidewalk so that Washington residents could pass by undisturbed (if not unleafleted).

That afternoon, Chairperson Hendrie and Commissioner Kennedy each made appearances on the street. The pale, gaunt Hendrie had disqualified himself from the Seabrook case because of previous conflicts; the stocky, effusive Kennedy was already known to be a staunch nuclear advocate, and had no fear of any possible legal action stemming from his talking with nuclear opponents in an "unbalanced" manner.

After shaking hands and answering questions on the street, Kennedy and Hendrie met with two Guard groups for more than two hours. Reviews were mixed, but many of the antinukers felt the two commissioners had spent too much of the time defending the industry they were supposed to regulate.

Ironically, the two commissioners known to be leaning towards suspension—Gilinsky and Bradford—asked to meet with the occupiers *after* the Seabrook decision. Their worry was that such a meeting prior to the decision might open them up to lawsuits if they found in the antinukers' favor.

The decision was due on Friday. On Thursday, fifty-six occupiers staged a "die-in" at the NRC doorstep. Taking the role of the commissioners, they approved a nuke, and then it melted down, "killing" them all. Because the "corpses" were blocking the NRC doorway, they were arrested, while the rest of the occupiers sang and chanted in support, and then applauded the D.C. police who handled the arrests without incident or injury, and who also seemed quite stunned by the applause.

The fifty-six were charged with "incommoding," held overnight in the legendary D.C. jails, and released Friday morning on $10 forfeit bond.

On Friday, with the first NRC arrests now history, both the heat and the tension mounted. While the occupiers waited, rumors about the decision abounded. Meanwhile, eighteen New Hampshire residents slipped into the NRC document room (which is open to the public) and vowed to stay. If construction at the Seabrook site was not suspended, they would repeat the original occupation of August 1, 1976, but this time inside the agency that had allowed construction to go ahead in the first place.

The decision was due at three. At three twenty, apologetic NRC staffers promised it by five. At five twenty, still no word. Then six. Then seven. Then eight. Nervous officials paced the NRC lobby while outside the occupiers geared up to shut the building.

Finally, at eight twenty it came. Commissioners Bradford and Gilinsky had outvoted Kennedy. Construction at Seabrook would halt as of July 21, pending EPA approval of the cooling system and a further inspection of alternative sites.

While the eighteen inside occupiers staged a victorious press conference, outside the Natural Guard treated Northwest H Street to an hour of joyous singing, dancing and hugging.

Once again, the Seabrook nuke had been shut down!

☆☆☆

The shutdown at Seabrook lasted only three weeks, while Douglas Costle scurried to fulfill the technical requirements needed to validate the EPA permit. But the June 24 rally and subsequent construction halt solidified conservative opposition to the plant. The feud between Thomson and Loeb, the miniscule turnout at the pronuclear rally in Manchester and the apparently magical ability of the Clamshell to turn disaster into victory all chipped away at the foundations of the Seabrook project.

The presence of major labor speakers at the rally also marked a particularly important turning point. (See Chapter Ten.) Joe Frantz of the United Steelworkers District 31 of Gary, Indiana, told the crowd: "We think the facts will show that the only way to provide reliable power, and the only way to provide more jobs in providing that power, is *not* to go the route of nuclear power.

"We know that there are more jobs provided by virtually every other route. Huge corporations are committed to nuclear power and they've thrown their financial power behind it. We don't have that money.

"But we have something more important. We represent the interests of the people of this country. And if we organize them, we'll have more power than those corporations will.

"That's why the environmental movement needs the labor movement, and the labor movement needs the environmental movement to achieve our ends."

Soon after the rally, former Governor Wesley Powell, of Hampton Falls, declared his gubernatorial candidacy. A conservative Republican with a strong personal following, Powell unleashed a vehement, unrestrained attack on Governor Thomson and the Seabrook project. Despite being vastly outspent, he carried 40 percent of the primary vote.

Encouraged by his showing, Powell decided to enter the November race as an independent. His unexpectedly high vote count led many New Hampshirites to believe that Mel Thomson was finally vulnerable—and they were right.

Around the United States, the nuclear industry was on the defensive, and for the first time, it showed at the polls:

Power at the Polls
(*The Valley Advocate*, December 6, 1978)

THE MAJOR MEDIA were pretty quiet about it, but the big loser in the November elections might well have been atomic power.

For the first time, nuclear opponents won a state-wide referendum victory, this one resulting in a virtual ban on reactors in Montana. A new constitutional amendment will do much the same in Hawaii, and a rate reform bill passed in Oregon will make future reactor construction there difficult at best. One Massachusetts district passed a resolution opposing further nuclear construction.

But the biggest blow of all came in New Hampshire, where Meldrim Thomson, godfather of the Seabrook nuke, was booted out of office on the issue of soaring electric rates.

Electoral Blow-Down

Up to this point nuclear opponents have had a sparse electoral track record. But in the fall a reversal of the momentum in nuclear balloting was registered all over the country.

Voters in the Twenty-seventh Middlesex, Massachusetts, state representative district (Cambridge) approved by an eight-to-three margin a resolution opposing further nuclear construction until a solution is found to radioactive waste disposal.

A Hawaiian constitutional convention (the state holds one every ten years) easily approved a constitutional amendment requiring two-thirds approval of the legislature before locating any plants or waste disposal sites. The state has no reactors now and the amendment, which was backed by the American Friends Service Committee and the local antinuclear alliance, seems to guarantee there won't be any in the near future.

In Oregon, a bill outlawing CWIP had been blocked in the state legislature. But organizers easily collected enough signatures to put it to a state-wide vote. Voters then banned CWIP with an overwhelming 68.1 percent of the ballot.

The biggest referendum surprise of all, however, came in Montana, where an antinuclear proposition had been soundly defeated just two years ago. Angered at having been outspent and out maneuvered in 1976, nuclear opponents were able to pull together a broad coalition that included ranking Democrats (including former U.S. Senate Majority Leader Mike Mansfield), key clergy, the state Small Business Association, the local Farmers' Union and the National Taxpayers Union for a second shot at limiting nuclear construction.

The new bill, called Initiative 80, required nuclear builders to post a bond equaling at least 30 percent of the ultimate construction costs. It also ordered them to prove that radioactive materials "can be contained with no reasonable chance of escape" and to prove the efficacy of the safety systems with "comprehensive testing of similar physical systems in actual operation." The bill lifted liability limits in case of a major accident, and required site approval by a state board of natural resources and by a "majority of Montana voters in an election called by initiative or referendum."

As usual, utility and industry contributers poured in funds to beat the bill. Among them was the Public Service Company of New Hampshire, whose $1,000 check was reported to the state election commission as having arrived on October 5. In all, nuclear supporters pumped in a minimum of $216,000, as opposed to $12,000 reported spent by the initiative's backers.

Industry ads pointedly labeled the the initiative a "ban" on nuclear construction, but its supporters won a Montana Supreme Court order barring those advertisements.

And, on November 7, those supporters won all the marbles, with Montana voters approving Initiative 80 by 159,035 to 87,876. Missoula County, which did have an outright ban against nuclear construction on the ballot, overwhelmingly approved it. "People in Montana don't like being told how to vote by outsiders," said a jubilant Jim Barngrover, a goat farmer instrumental in the campaign. "Especially outsiders with vested interests."

Showdown Over Seabrook

And while Montana organizers were jubilant, activists in New Hampshire were positively ecstatic.

Since 1972, Republican Governor Meldrim Thomson has ruled the state as if it were a fiefdom. In league with William Loeb, Thomson dominated the state with an iron hand. To highlight his right-wing beliefs and national ambitions, Thomson traveled to South Africa in support of apartheid. He ordered state flags to half-mast when Taiwan was dropped from the Olympics. He barred from driving in New Hampshire a Massachusetts driver who had gestured obscenely at the Governor's car.

Though the soft-spoken Georgia native often appeared as a media buffoon, his antics inside New Hampshire were not to be taken lightly. He dominated police, patronage, the media, and the courts and, in league with Loeb, created an atmosphere that could be moderately described as "oppressive."

But in the summer of 1978 Thomson began having serious problems. When the Clamshell Alliance agreed to stage a legal rally on part of the Seabrook site set aside by state Attorney General Thomas Rath, the *Union-Leader* had screamed for blood. The Governor, the paper charged, was weak and even "a liar." Though Loeb and Thomson soon reaffirmed their mutual support, the bitter, highly visible split shook the state.

Thomson also found himself in the position of having to support higher electric bills. His main source of political strength had been his pledge to oppose all state income and sales taxes. In his three prior races he had beaten opponents who couldn't match his antitax zeal.

This fall, however, his support for Seabrook got in the way. Badly strapped for cash to build the mammoth twin reactors (cost currently estimated at $2.5 billion), the Public Service Company of New Hampshire was forced to raise electric bills by 17 percent through the same kind of CWIP arrangements that were rejected by voters in Missouri and Oregon. Rates, said the PSNH, would have to go up as much as 50 percent by 1982, when Seabrook was slated to go online.

Prior to the summer Clamshell rally, the New Hampshire legislature voted to ban CWIP, a move the PSNH charged would kill Seabrook and probably the company as well. Thomson put his neck on the line by vetoing the ban. The legislature was unable to override.

But Thomson's Democratic opponent, Nashua car salesman Hugh Gallen, now had his opening. Taking the pledge against sales and income taxes, Gallen launched an attack against CWIP, which he branded "Thomson's Tax." Thomson tried to outflank Gallen with right-wing issues such as the death penalty. But Gallen used Boston television to skirt the *Union-Leader* in southern New Hampshire, and managed to keep it a single-issue contest, pledging his support for Seabrook but vowing, if elected, to axe CWIP.

A wild card named Wesley Powell also confused the picture. A fundamentalist conservative from Hampton Falls, which borders on Seabrook, former Governor Powell had run a bitter anti-Thomson primary, charging him with chicanery, buffoonery, and sins against the church in Loebesque terms. After a respectable but losing showing, Powell leapt into the main round as an independent.

Democrats claimed Powell was draining their support. But, in fact, Powell's ferocity added strength to the anti-Thomson attack as he slammed CWIP, the Seabrook plant, and Meldrim with unyielding vigor.

Early election night it appeared that Thomson had weathered the storm. His supporters began celebrating.

But as the evening wore on, the results turned around. By morning, Hugh Gallen was the clear victor, claiming a 50-46 percent margin. Powell carried some 12,500 votes, just a bit more than the margin between Gallen and Thomson. Moderates throughout the state expressed joy and relief hard for outsiders to appreciate. "The reign of terror is finally over," said Dan Keller, a state native from Laconia.

A Gallen of Nukes?

Gallen's election has cast a long shadow over the Seabrook project. Though he has reaffirmed his support for the plant, a meeting with PSNH President William Tallman left little doubt about the fate of CWIP. The company, said Tallman, was giving up on CWIP. "Some other way" would have to be found to finance the plant.

Where the search leads will be watched with great interest. Within hours of Thomson's defeat, PSNH announced it was "temporarily postponing" a two million-share offering of common stock originally planned for November 14. PSNH Vice President Robert Harrison last week reportedly confirmed that the company was having "serious . . .short-term cash problems." According to wire service reports, the company has used up most of a $100 million line of credit with major Boston and New York banks, and now faces exhaustion of its ability to borrow. The cancelled stock issue was to have temporarily filled the gap, but, without it, says Harrison, construction could come to a halt within weeks.

The impact of Thomson's defeat—and of another potential shutdown—on the world's best-known reactor project is obviously enormous. The new governor is publicly committed to the plant's completion. Whether he'll go out and stump for it is altogether another question.

But some key New Hampshire strongholds—particularly the governor's council, which approves state appointments—already reflect a growing opposition to the plant. Whether the plant can survive the four years remaining until its scheduled opening is at very least a $64 million dollar question.

☆☆☆

Two days after Thomson's defeat, the PSNH withdrew a $40 million stock offering. A bond sale earlier in the year had gone without takers. The PSNH cash-flow situation was critical.

Now, said the company, everything would hinge on the outcome of CWIP negotiations with the new governor.

But as Hugh Gallen moved into office, he made it clear there was nothing to negotiate. He supported the Seabrook project, but he opposed CWIP. Within months both the House and Senate voted to ban it. This time, there would be no gubernatorial veto to override.

Meanwhile the utility announced that even with CWIP it would have to sell a major portion of its holdings in the plant. At least 60 percent of its 50 percent share would go on the block— leaving just one fifth of the state's biggest construction project in New Hampshire hands.

The likeliest buyer for PSNH's excess shares seemed to be the Massachusetts Municipal Wholesale Electric Company (MMWEC), a consortium of twenty-nine Massachusetts community electric companies. The purchase was strongly supported by Edward J. King, the new pronuclear governor of Massachusetts.

The move also marked an important shift in atomic financing. Like several other utilities committed to reactors, the PSNH was now turning to publicly-owned companies to support private investments. Because it represented municipal utilities, MMWEC was permitted to float tax-free bonds. It could also avoid bankruptcy in a pinch by automatically passing on cost overruns to its customer-owners.

As the PSNH kept construction at the site going with a $100 million line of credit from Boston and New York bankers, the municipalities in line for the extra shares geared up for some fierce debate.

Meanwhile 2900 workers reported daily to the Seabrook site. By mid-1979, the plant was 15 percent complete.

Clam Waves

Following the June 24 rally, a series of "wave actions" kept civil disobedience alive at the plant gates. Among those arrested was Benjamin Spock, the baby doctor. "We come here with a very high purpose," Spock had told the rally on June 24. "We've got to defend people, millions of people, not only in this region, but people in all other regions of the U.S., who've got to be inspired by the success that's achieved here. Most of all, I feel a responsibility of defending the children. The adults, if they allow themselves to be incinerated or otherwise poisoned by nuclear power, in a sense it will be because they've been too inert to rise up and fight and demand their rights. Children can't do that, and we have to speak for them."

In March, 1979, the PSNH brought in the 450-ton reactor pressure vessel (RPV) for Seabrook Unit 1. The Clamshell had been preparing a sea blockade against the RPV, which was in storage at Fall River, Massachusetts. But the company made an end run, bringing in the RPV for Unit 2 from Chattanooga, Tennessee, instead. PSNH representative Norm Cullerot said the two were identical, and that the company had secretly ordered up RPV 2 to avoid the Clamshell blockade. "Isn't that too bad!" he gloated.

"It's pretty amazing," replied organizer Kristie Conrad in an Alliance press statement, "when the largest utility in New Hampshire has to sneak in a 450-ton piece of equipment just to avoid a citizens' group."

On March 9 the Clamshell regrouped and staged a series of actions against the moving of the RPV overland through Seabrook, resulting in some 190 arrests.

But by then Alliance members were involved in a serious rethinking of tactics, strategies, and goals. The Clamshell had survived more than thirty months of intense, successful organizing. At least three-dozen parallel alliances had sprung up around the United States to stage organizing campaigns and nonviolent actions at scores of reactor, uranium mining, weapons, waste storage, and fuel fabrication centers around the country. The thirty-five persons who had founded the Alliance could now point to at least twice that many safe-energy groups around New England, each with its own distinct character and each firmly rooted in the community from whence it came. Many were now deeply involved in the work of physically converting their locales to natural energy sources.

Shortly after the blockade, the Clamshell held a weekend congress in Worcester, Massachusetts, to pull together the loose ends that had frayed since June 24. Its decision-making process, its reliance on civil disobedience, and its cohesion as a group— all came up for serious rethinking.

But the ongoing fight at Seabrook remained. Despite two regulatory shutdowns, the defeat of Meldrim Thomson, the tripling of the cost of the plant, the near-bankruptcy of the PSNH, the flight of a long string of utility backers, and opinion polls that showed a majority of New Hampshirites opposed its construction—the Seabrook nuclear power plant continued to be built.

Few opponents of the project harbored any illusions. The stakes at Seabrook had become the very highest, not just for the nuclear industry, but for the way decisions are made in a corporate society. Could the United States of the 1980s allow a citizens' movement to shut down an industrial project that had become a prime symbol not only of the world energy war, but for grass-roots citizens' campaigns everywhere?

To what lengths would government and industry go to keep the PSNH out of bankruptcy and the Seabrook project afloat? What price would they pay, what toll would they take, to keep the reins of decision-making power?

Conversely, if the very best efforts at grass-roots and electoral campaigning failed, how could the Seabrook nuke be stopped? To tens of thousands of New Englanders, the idea of that plant's going online was now absolutely unthinkable.

If the time came to fire the reactors at Seabrook, would a mass occupation become the movement's last resort? Was it realistic? How many nonviolent people would it actually take to stop a nuclear reactor from operating? Would a hundred eighty thousand do it? Half a million? A million? How would they all be trained? How would nonviolent discipline be maintained? What, indeed, were the social implications of shutting down an industrial project in which so much financial and political capital had been invested?

As an umbrella coalition, the Clamshell Alliance had learned that the fight for decentralized, democratic energy would be won first and foremost in the neighborhoods and communities where energy efficiency would be improved, utility bills resisted, and renewable power generated. Centralizing organizations could come and go; it was people at the grass roots who would finally win the war.

In the course of the weekend congress at Worcester, the Clamshell abolished its consensus process, developed a summer program, and moved to patch up its internal differences. "The company can't finish that plant until 1983," said Robin Read on March 19, the last day of the congress. "If they get that far, we'll be ready."

Nine days later a series of malfunctions would strike a nuclear power plant in Pennsylvania. Three Mile Island was just around the corner.

6
Shoot the Devil: California and Other Struggles

☆☆☆

The battle over atomic energy that was beginning to dominate New England had also emerged elsewhere in the United States, perhaps nowhere more markedly than in California.

In 1976 the state had held the nation's first big referendum on atomic power. The industry won at the polls, but not before conceding three state laws limiting nuclear expansion in the state and making further construction dependent on a solution to the waste disposal problem.

Nuclear power in California faced other special problems—notably earthquakes. After years of legal battle, a reactor project at Bodega Head became one of the first in the United States to be scrapped in midstream when intervenors proved the site was directly on a fault line. Later a reactor at Humboldt Bay was shut down after fourteen years' operation because it, too, sat on a seismic fault.

No such problem seemed to exist in Kern County. The rich, inland agricultural area was very conservative, had voted better than two to one against the antinuclear Proposition 15, and seemed safe ground for nuclear expansion.

But voting for atomic power in general is one thing; voting for plants in one's own back yard is quite another. The Los Angeles Department of Water and Power (LADWP) had failed to account for things like water supplies and local fears over steam from cooling towers that might unbalance weather patterns.

Such "non-nuclear" issues were becoming prominent at proposed reactor sites around the country. In Kern County, in a vote that may have marked the end of nuclear development in the nation's largest state, they proved fatal:

Nuclear Defeat in Kern County
(Pacific News Service, March 11, 1978)

Bakersfield, California

THE OVERWHELMING REJECTION of a nuclear power complex by voters in this conservative agricultural community has underscored a growing national trend among rural areas to oppose atomic reactors for essentially non-nuclear reasons. And it has cast a long shadow over the future of atomic energy in California—while strengthening the increasingly prosolar hand of Governor Jerry Brown.

Voters in Kern County in south-central California's San Joaquin Valley, on March 7, decisively rejected plans for a mammoth 5200-megawatt reactor complex slated for the town of Wasco. The vote was the first popular vote on atomic energy in California since the June, 1976, rejection of a referendum aimed at restricting the state's nuclear development.

A staunchly conservative farm area, Kern County registered a strong pronuclear vote in that 1976 referendum. But when it came to allowing nuclear facilities in their own back yard, the county made a stunning turnaround, rejecting the measure by a margin of almost five to two.

A wide variety of local issues led to that rejection. Much opposition stemmed from the fact that the nuclear power plant was being spearheaded by the Los Angeles Department of Water and Power, and that the power was essentially for Los Angeles. Less than 10 percent of the project's electricity was slated for use in Kern County.

The idea of a "Los Angeles nuclear power plant" in Kern County stirred resentment and angry memories. Back in the twenties Los

Angeles developers turned to Kern County for water and engineered a dubious deal that served as the basis for the movie "Chinatown." Bumper stickers reading "Remember Owens Valley" are still common on Kern County pickups, and residents did not welcome another intrusion by a Los Angeles utility.

The battle also reverted to the old question of water. The enormous reactors were scheduled to consume thousands of gallons for cooling, and after three years of drought many local residents were dubious about where the operators would get it.

Officially, the plant was to be cooled with runoff from agricultural irrigation. San Joaquin farmers use deep irrigation trenches to capture runoff, and the plant builders hoped to use this to cool the reactor.

Under pressure, they also quietly admitted contingency plans to tap the California aqueduct, which supplies much of southern California's drinking water.

Local farmers greeted the plan with skepticism, speculating that the reactors might deprive them of precious irrigation water. The DWP also planned to flood 35,000 acres of farmland for a reservoir. Since the irrigation runoff would be saline in content, it would need to sit idle for a time while particulates settled out and the water became fit for cooling the reactors. The prospect of losing nearly six square miles of farmland angered locals, who claimed it would cost the region thousands of agricultural jobs and substantial income.

Farmers also feared unbalanced weather patterns caused by steam emissions from the plant's huge cooling towers. Because the steam would also be chemically treated to keep it free of micro-organisms, farmers were concerned about chemicals settling on their fields, as well as damage to local weather patterns.

Indeed, the well-organized campaign to defeat the measure became one of the few in which both farm owners and workers have officially joined hands. "All in all," twenty-six year old Wasco farmer Jeff Fabbri said just before the vote, "I can't really think of any good reason to vote for the damn thing."

Nonetheless, the overwhelming rejection of the plant came as a shock to nuclear supporters who poured thousands of dollars into the county in a pre-election blitz.

The Kern County vote has already had a significant impact on future energy policy in the state of California.

Nuclear power has become one of the key campaign issues in the re-election effort of Governor Jerry Brown. Brown has become increasingly antinuclear in recent months, terming the issue "the next Vietnam" and pledging to veto a bill now making its way through the legislature that would allow construction of a reactor complex at Sundesert near the state's Arizona border.

Brown's Republican opponents challenge the Governor's "windmill and woodchip" energy program and seem intent on making atomic power the chief issue of the campaign.

"A rejection of nukes by such conservative areas can't help but strengthen Brown's hand," says long-time California antinuclear activist John Berger. "The Republicans have to be wary now of taking a strong pronuclear stand."

Indeed, Brown, who had already pulled the state out of a 10 percent share in the San Joaquin plant, greeted the Kern County vote as "a message to those who spent millions of dollars on nuclear research" that they had better look at alternative sources of energy.

☆☆☆

Despite his "windmill and woodchip" energy policy, or possibly because of it, Brown won re-election easily in 1978. Pronuclear legislation designed to pave the way for construction at Sundesert soon failed in committee, and the nation's most populous state found itself with a moratorium on new nuclear construction. Two reactors, Rancho Seco near Sacramento and San Onofre I near Los Angeles, remained in operation. Four more were being built, two at the San Onofre site, and two more at a place called Diablo Canyon, near San Luis Obispo. The two at Diablo would soon be dubbed "Seabrook West".

The Tremors at Diablo
(*The Progressive*, April, 1979)

IN THE COASTAL HILLS OF CENTRAL CALIFORNIA, fifteen miles south of San Luis Obispo, stands a landmark in the national energy debate—the bitterly contested Diablo Canyon nuclear power plant.

In recent months, Diablo has become the focus for a wide range of key national regulatory questions. And it has put California's ambitious Governor Jerry Brown on the hot seat with the burgeoning grass-roots antinuclear movement, a force that may have to be reckoned with in the 1980 Presidential campaign.

Diablo's unique location is what has made it so crucial. The power plant is almost astride an earthquake fault. The site was selected by Pacific Gas & Electric (PG&E), the nation's second-largest private utility, in the fall of 1966. Two Diablo units, totaling 2,212 megawatts, were to be finished in the mid-1970s at a cost of $350 million.

But PG&E ran into delays and cost overruns of the type that have plagued the nuclear industry everywhere. People in the San Luis Obispo area raised such issues as evacuation, potental sabotage, and general plant security. Faulty welds and inferior fuel rods delivered to the site also were matters of concern.

In 1971 there came a bombshell from an unexpected source—Shell Oil Company. Two Shell geologists prospecting off the coast discovered a major earthquake fault just 2.5 miles from the Diablo site. The work went on, and PG&E now claims it learned of the discovery much later through a magazine article. The Nuclear Regulatory Commission (NRC) was not informed until 1973.

Discovery of what became known as the Hosgri fault created a sensation, especially when the U.S. Geological Survey confirmed that it was capable of a 7.5 Richter-scale quake—ten times the force Diablo had been designed to withstand. San Luis Obispans quickly asked the NRC to stop construction at Diablo until full ramifications could be investigated. In 1974, the Atomic Safety and Licensing Board (ASLB) refused that request, but assured intervenors that the hundreds of millions of dollars PG&E was pouring into the site would have "no bearing" on whether the reactors should be licensed.

Five years later the battle still rages and the question of Diablo's operating license has taken on national proportions. Seismic questions have become key factors in reactor litigation on both coasts, and the Diablo decision may be considered a precedent-setter. "If they can license this one," says one intervenor, "then they can license anything."

Diablo has acquired additional importance because there is strong evidence that the NRC staff may have "fudged" opinions on the case for political reasons. According to internal documents made public through the intervenors, NRC researchers wrote at least one 1976 memo in which politics clearly outweighed engineering: "The impact of potential denial for operation" at Diablo, they said, had to be weighed against "the viability of continued operation of plants at other sites." In sum, said the staff, there was a natural reluctance to deny PG&E's license "because of the large financial loss involved and the severe impact such action would have on the nuclear industry." Among other things, the NRC suggested a unique "interim" license that would lessen that impact by paving the way for the plant to open on a "temporary" basis.

The discovery of the earthquake fault did prompt PG&E to "retrofit" the reactors for additional strength. But intervenor attorney David Fleischaker says "the bulk of the effort has gone into explaining away the potential effects of the earthquake" rather than building to withstand one.

Fleischaker's charges are buttressed by another leaked document, a July, 1978, staff letter to NRC Chairman Joseph Hendrie admitting that the plant retrofits could not meet the same standards expected from a project originally designed to withstand a shock of the magnitude possible at the Hosgri location.

Intervenor reaction to that memo was especially strong. Other evidence had already surfaced that PG&E's own consultants knew as early as 1967 that the offshore area was prime earthquake territory and that further exploration was needed. But when Ralph Vrana, a geologist from nearby California Polytechnic Institute, made just such a suggestion public, he found himself without a job.

Meanwhile, PG&E has sunk nearly five times the original cost estimate into the plant. "Diablo raises some really fundamental questions," says Raye Fleming, one of the plant's outspoken opponents, "like how close can they build to an earthquake fault and then blackmail the public into accepting it because of money spent? And then, how much of that can they turn around and force the public to pay for?"

Indeed, throwing the switch at Diablo could hand the California Public Utilities Commission its toughest case in years. The Commission must determine how much of Diablo's final price tag should be charged to ratepayers and how much should come out of PG&E's profits as a penalty for company mismanagement. The California decision will be watched closely by utilities and nuclear opponents everywhere; the question of who is to pay for the more than $1 billion in cost overruns at Diablo could have a major impact on the kind of risks utilities will be willing to take to build future atomic plants.

Diablo's fallout could also be crucial for the ambitions of Governor Brown.

In the spring of 1978, voters in conservative Kern County overwhelmingly rejected a major reactor complex proposed for the town of Wasco. The vote was considered a turning point; it led to the legislative rejection of a bill designed to allow construction of another nuclear facility at Sundesert, near the Arizona border.

Together the two decisions gave California a *de facto* moratorium on new nuclear construction and prompted Brown to take a highly visible antinuclear position in his successful 1978 re-election campaign against Eville Younger, a pronuclear Republican.

But Brown may now face some ticklish decisions about Diablo. Will he be content to let the plant start generating power? What will he do about those who are determined that it shall not?

More than five hundred demonstrators have already been arrested for Diablo "occupations" organized by the Abalone Alliance, a California coalition of nuclear opponents. Court handling of the resulting criminal trespass trials has brought an angry reaction. San Luis Obispo judges have been handing down long jail terms and heavy fines, despite the fact that no violence or property damage has occurred. Normal penalties for criminal trespass are a couple of days in jail and minimal fines.

In mid-December, Municipal Judge Robert Carter also refused to allow Dr. John Gofman, an authority on the health effects of radiation, to testify to a jury on the dangers of atomic energy. The Abalone Alliance had hoped to "put nuclear power on trial" by telling the jury why the twenty occupiers on trial had felt compelled to break the law in order to stop Diablo. But the judge ruled discussion of the issue "irrelevant" to the trespass charge, and the jury found the defendants guilty. Some of the jurors wept openly as the verdict was read, and the foreman later asked for leniency in sentencing. But Judge Carter was not moved. He handed down sentences of ninety days in jail and $400 fines. Although Carter later apologized to the "Diablo 20" and lowered their sentences, the judgement seems to have heightened the confrontation atmosphere surrounding the plant and toughened the resolve of many in San Luis Obispo.

"Too many people have worked too long and too hard against that thing," says Raye Fleming, "and too many people are getting seriously worried. I just can't imagine how people will react if they throw that switch."

☆☆☆

After Three Mile Island Governor Brown moved into an increasingly antinuclear—and national—posture. He wrote the Nuclear Regulatory Commission asking that Diablo Canyon be kept shut until a full review could be done of the Three Mile Island accident, and pending a full study of Diablo's ability to withstand seismic shock. He also asked that the licensing be held off until the state could update its "emergency preparedness measures."

Brown's letter to the NRC came just two days before his appearance, with Jane Fonda and Tom Hayden, at a May 6 protest rally in Washington, D.C., which drew an estimated 125,000 participants, the largest crowd in the movement's history.

Meanwhile the pressure to keep Diablo Canyon shut reached the boiling point. On April 7 some forty thousand people swarmed into a "Stop Diablo" rally at San Francisco's Civic Center. Thousands more gathered in the summer near San Luis Obispo. And the Abalone Alliance continued in its vow to block operation of the plant if and when it got its license, a threat they now seemed able to back up with the promise of thousands of occupiers. With $1.4 billion at stake, Diablo will be at the center of the nuclear struggle for years to come.

While the fight over Diablo escalated, so did the level of confrontation at nuclear sites around the country. But nowhere was the "energy war" more brutal, or the stakes higher, than in a battle over high-voltage power lines being waged by a unique coalition of farmers, Indians, and urban activists in Minnesota:

Farmers on the Line
(*The Progressive*, July, 1979)

Lowry, Minnesota

THIS FLAT, FERTILE, MIDDLE-AMERICAN FARMLAND has become a battlefield in the most raucous and explosive "energy war" in the United States, one that has turned some of America's most conservative farmers into angry eco-raiders, garnered the support of half the state for their guerrilla tactics, and blasted the powerful Democratic Farm-Labor (DFL) party into a heap. It may also be serving as the vanguard for the most critical fight of the 1980s, the struggle for solar power.

The three-year-old battle is over high-voltage power lines—the biggest in the United States—designed to carry 800 kilovolts of direct current from an 1100-megawatt coal-fired generator at Underwood, North Dakota, to a switching station near Minneapolis. Construction has destroyed some eighty-five hundred acres of prime farmland, and the farmers not only don't like it, they don't intend to allow the power line to operate.

Last summer, for example, about one hundred fifty residents of Stearnes and Pope Counties, many of whom regularly milk thirty or more Holsteins, held a wienie roast near a power-line construction site. But the festive atmosphere was shattered when eighteen patrol cars roared up to the bonfire. Heavily armed state troopers and sheriff's deputies leapt out of the cruisers and ordered the party to pack up.

The farmers had other ideas. First they blinded the police with high-powered lights. Then they opened fire with makeshift slingshots known as "wrist-rockets." "I'd say there wasn't a single window left on those cop cars," recalls one local resident with a chuckle. "Those cops never knew what hit 'em."

Angered by the incident, some of the farmers disappeared into the fields, and within hours a "natural plague" began afflicting the line. A swarm of what the farmers call "Bolt Weevils" (*weevulus unboltus*) assaulted many of the one hundred fifty-foot-high, $80,000 towers, causing them to fall. Hordes of "giant blackbirds" spat shotgun pellets into the glass insulators that prevent the power line from short-circuiting, doing about $5000 damage per blast. By the end of the night, large segments of the line had been mortally afflicted.

Ironically, many of the power line's staunchest opponents are part owners of the companies building it. The Minnesota line is the long tail of a $1.25-billion project that includes a lignite strip mine, North Dakota's Underwood coal station, the power line itself, and a switching station at Delano, near Minneapolis. The project is owned by two rural co-operatives, the United Power Association (UPA) and the Cooperative Power Association (CPA), which serve much of the area through which the line passes. Farmers argue bitterly that the management of the co-ops has been taken over by urban-based private utilities and is beyond their control.

And at any rate, the project is not for the farmers. "The whole thing is being built to serve the Midwestern power grid," says John Kearney, a member of a Twin Cities antinuclear and antipower-line group called Northern Sun Alliance. "The electricity will go to the cities and the profit will go to the big private utilities!"

Indeed, if the costs of the project are added to their rate base as planned, farmers along the line—who will get no power from it—can expect their electric bills to triple.

And if the line becomes operative, they can also expect some severe side effects. "They can't prove it's safe," says Kenny Thurk, a farmer from Villard, Minnesota, "and they can't prove they need it."

Thurk, thirty-eight, has snapshots of himself being dragged through the snow and thrown over a fence by state troopers. "Knee'd me in the back and jumped on my kidneys," he says wryly. "Never been arrested before in my life. Jackie neither," he adds, nodding toward his dark-haired wife, who was recently convicted on charges of aggravated assault stemming from an attempt to block a cement truck headed for the power line.

Shortly after her conviction, a tower near the Thurk farm crashed to the ground, debilitating the line. "They haven't had any experience with that much direct current," Jackie explains. "We don't know what it can do to us."

The power co-ops have told the farmers that the line is perfectly safe. But a number of recent studies have linked high-voltage lines to stunted growth, birth defects, heart disease, mental disorders, sexual impotency, and cancer. For farmers with first-hand experience, other damage is irrefutable—decreased fertility and lowered milk production among cows, imperiled crop production, and heavy shocks delivered to farmers working with metal equipment, especially during rainstorms or periods of high humidity. "Those damn lines, on a wet day, can throw shocks all over the place," says Kenny Thurk. "We've had reports from farmers in Ohio that they can't even drive near the thing without worrying about their hair standing on end. If you want to work near the lines, everything you've got has to be grounded, including you."

Power-line operators also routinely spray the right-of-way with toxic herbicides, a practice with serious health implications for anyone living or eating food grown nearby. Dr. Andrew A. Marino, a research physicist at the Veterans Administration Hospital in Syracuse, New York, recently told the New York Public Service Commission that "every biological effect induced in people by overhead high-voltage lines is potentially hazardous and should be avoided." At best, he added, the insertion of such lines is "an enormous, utility-operated human experimentation program."

Small wonder the farmers were hesitant to sign up. In 1976, small groups of them began banding together to block surveyors from taking their readings. "The companies just came in here and started pushing people around," recalls Gloria Woida of Sauk Centre. "They said, 'We're going to cram this power line down your throats.' "

The farmers decided to cram back. As opposition and sabotage escalated, Alice Tripp, a feisty retired farmer from El Rosa, decided to enter the 1978 DFL primary for governor. Spending only $5000 on her campaign, she stomped the rural areas, promoting the fight against the line. "If I were any good with a wrench," she declared, "I'd be out there unbolting towers myself." The pollsters and media scoffed, but when the dust settled, Ms. Tripp confounded the experts by garnering 100,000 votes, 20 percent of the total.

That wasn't the end of it. Most observers believed the DFL would carry the governorship and both U.S. Senate seats, which were up for grabs that fall. The DFL had dominated Minnesota politics for thirty years, and despite its growing problems, the legacy of Hubert Humphrey seemed strong enough to carry the party to one more victory. The DFL has long been the key Democratic stronghold in the upper Midwest, and just to make sure it held, both Jimmy Carter and Walter Mondale made special campaign trips to Minnesota, underlining their view of it as a "must-win" state.

Both incumbents—Wendell Anderson, who was running to hold his Senate seat, and Rudy Perpich, the incumbent governor—had run roughshod over the farmers. There were extenuating circumstances. Anderson had been elected governor, but alienated many voters by appointing himself to the Senate seat, and thus may have made things more difficult for Perpich, whom he had appointed governor.

But ultimately it seems to have been the farmers who made the difference by either voting Republican or not at all. Perpich lost by roughly the same number of votes Alice Tripp had drawn in the primary, and Anderson was crushed in his race for Senate. The DFL lost the other Senate seat as well, largely because Don Fraser, its strongest candidate, who was also distrusted by the farmers, had failed to survive the primary. "The national media ignored it," says John Kearney, "but it was the power line that beat the DFL. They could have won with the farmers. Instead they sent in the state troopers. Now they're out."

A state-wide poll conducted early in 1978 by the *Minneapolis Tribune* showed 63 percent of the public opposed to the power line, and 50 percent overall supported breaking the law to stop it.

But despite having turned Minnesota politics upside down, most of the farmers had little or no political involvement prior to the arrival of the power line. Many who vehemently supported the war in Vietnam now welcome scruffy veterans of the protests into their ranks. "A lot of what the so-called city radicals have been saying has been right on the head," says Gloria Woida, a thirty-eight-year-old mother of five. "To me they're like family."

"We don't care how they look or what they did in the past," adds Alice Tripp, on whose farm six Twin City activists were recently arrested. "It's time to forget our differences and work together to stop the corporations and bureaucracies from rolling over us."

The farmers' General Assembly Against the Powerline (GASP) has also made strong connections with the NativeAmerican movement. When felony charges were brought against the farmers (to date nearly two hundred arrests have been made in some seventy separate incidents), they recruited Twin Cities lawyers from the Wounded Knee Legal Defense/Offense Team. At a recent GASP rally, Native American activist Clyde Bellecourt called the farmers "the new Indians," and a farmer-Native American-antinuclear organizing meeting in Rapid City, North Dakota, in March set a year-long joint program for antimining and antipower-line activities. Among other things, it will include a July march against uranium mining and a massive "Survival Fair" to take place in the Black Hills in the summer of 1980.

The coalition of farmers, Indians, and urban activists obviously goes way beyond personal affinities or legal defense strategies: it cuts to the heart of the American power system.

The Carter energy plan calls for a massive increase in the use of coal, much of it strip-mined from western native American lands. Underwood is one of the first in a mushrooming network of generators being built at the mouths of western mines. Its operators are vehemently antiunion. "We at the Falkirk Mining Company are deeply committed to the preservation of nonunion status," North American Coal Corporation's Robert Murray told a 1975 meeting of the Cooperative Power Association. "It is a positive effort of restructuring our employee's relations to increase productivity and motivation."

It is also, as Murray repeatedly told his audience, a crucial link in a stategy aimed at avoiding the power of the unions in Appalachian coal. "Coal by wire," that is, power generated by coal and shipped over high-voltage transmission lines, would avoid the rail costs of shipping coal, the environmentalist roadblocks against burning coal in urban areas, and the labor costs of dealing with the United Mine Workers.

But the farmers may have found the system's Achilles' Heel. "What they've done," says Kenny Thurk, "is build four hundred twenty miles of power lines they can't defend." Indeed, the CPA/UPA management recently pulled its armed guards off nightime duty, essentially conceding that the seventeen hundred towers and thousands of insulators can't be protected. At least five towers and hundreds of insulators have already been destroyed.

That crunch of metal, glass, and wire could translate into a deafening roar for America's energy planners. Without that transmission channel, there's no future for centralized power generation in the United States—coal, nuclear, or otherwise. Militant opposition to high voltage lines has already surfaced in New York's northern Hudson Valley, is developing in Alabama and New Hampshire, and is bound to spread.

Meanwhile, martial law has already been tried in Stearnes and Pope Counties, to no avail. The CPA/UPA operatives are now waging a classic war of attrition, buying out some farmers, threatening others with lawsuits and prosecution, hoping to open the line this summer.

But the companies are in serious financial trouble, and the farmers remain confident. The battle lines also seem to be hardening. "There's two kinds of power," says Gloria Woida. "Money power and people power. And I'll put my money on the people."

The recent nuclear accident at Three Mile Island, Pennsylvania, has also deepened the Minnesota farmers' opposition to atomic generators operating at Monticello and Prairie Island. Farmers in nearby western Wisconsin helped win the recent cancellation of a proposed plant at Tyrone, southeast of the Twin Cities.

Indeed, for many of the farmers, the fight against the power line has quickly translated into a fight for solar energy. Many now claim that the lignite coal mined for use at the Underwood generator is laden with enough radium and uranium traces to make it a significant radiation producer, and the farmers don't seem to like it much more than they like the nearby nukes. "What we really need," says Gloria Woida, "is to get our farmers self-sufficient. We don't need those damn nuclear plants or strip mining. We've got wind and methane, and we can grow our own grain for alcohol. If we can't get permits, we'll make the alcohol just like in the moonshine days. And I tell you something. If the people decide that solar energy is what we want, there's just no way the government is going to squash it."

☆☆☆

The Minnesota power-line war was matched in its effectiveness by a much quieter, but parallel struggle a thousand miles to the south, where a homogeneous Mexican-American community took control of its own energy supply:

An Energy Rebellion

(*The Progressive*, August, 1978)

Crystal City, Texas

THERE ARE DOZENS of community-owned cords of mesquite stacked at a shack on the outskirts of this dusty town of 8,100 in southern Texas. The shack itself is fitted with a solar greenhouse and serves as transfer point for Korean War surplus wood stoves, which have gone to about a thousand local households. The system of distributing fuel and stoves has become an important cog in this Chicano community's stunning attempt at energy independence.

Crystal City became something of a national symbol last September, when Lo-Vaca Gathering Company, a natural gas distributor, won a court ruling authorizing the quintupling of gas rates to its south Texas customers. The municipality of Crystal City (which residents call "creesTALL") had been acting as distributor for the town's fuel, but in 1975 the Texas Railway Commission, the state's attempt at public utilities regulation, allowed Lo-Vaca to raise gas rates from about thirty-six cents per thousand cubic feet to $2, despite long-term contracts with some two hundred Texas cities. By the fall of 1977, a court confirmed the increase. This foretaste of national deregulation left Texas consumers gasping—but only Crystal City defied the company and told it to keep its gas.

The reasons are not hard to find. For eight years, politics in Zavala County—and especially in Crystal City, the county seat—have been dominated by La Raza Unida, a Chicano party which seeks more than just energy independence. "This movement is the browning of America," says Jose Angel Gutierrez, a founder of La Raza Unida and now Zavala County judge. "It means giving hell to the owners of the restaurants on the other side. It means giving hell to the owners of Del Monte. It means taking all the land from the ranchers in this county. It means bettering ourselves, as Mexicans, as the bronze race, throughout the country."

In 1970, La Raza Unida took control of all five seats on the Crystal City council, and seven of ten on the town and county school boards. Two years later, the party ran candidates in state-wide elections, and did well enough to throw a scare into the Texas ruling establishment. La Raza's gubernatorial candidate, Ramsey Muniz, has since been charged with possession of marijuana, and, according to Gutierrez, more than a dozen Raza Unida activists have been jailed on trumped-up charges.

The party itself has become factionalized, but seems to have strengthened its hold on Crystal City. The gas crisis has not hurt. "Lo-Vaca did us a great favor by cutting off the gas," says Gutierrez. "It really gave the people a kick in the ass, forced us all to get moving."

When two years of court battles finally resulted in a company victory, Gutierrez, as county judge, declared Crystal City a disaster area, and called for federal funds. With the help of Senator Edward Kennedy, Community Services Administration (CSA) money came through to buy canisters of butane to help local residents through the winter.

But there were problems. The canisters were to be shared among households, making accountability difficult. The butane gas was also more expensive than Lo-Vaca gas. Nor was the supply renewable— soon the bills would start pouring in, and Lo-Vaca was still trying to get some $800,000 it claimed Crystal City owed in back payments.

Kennedy's help also infuriated Texas Democrats, who view La Raza as subversive. Governor Dolph Briscoe, the state's largest landowner, bases his fiefdom in nearby Uvalde, and has continually used his considerable influence to cut federal funds destined for his upstart neighbors. Under Raza Unida control, he charged, Crystal City had become "a little Cuba."

The charge is far from accurate. Though about 80 percent of the population consists of migrant farm families, Crystal's economy is still gringo-dominated. White-owned businesses line the paved main thoroughfare, while Chicano stores and cantinas overlook the dusty, unpaved streets of "Mexico Chica."

The area is known as the "winter garden" region of south Texas, because crops can be grown there year-round. Zavala County plays host to many multinational agribusiness corporations, one of which has adorned the area with a life-sized statue of Popeye in honor of the region's spinach crop. "Only thing the gringos ever gave this town," says Gutierrez.

Not far from the statue, however, is the community clinic, staffed by two doctors and a dentist whose fees for a visit are based on ability to pay, with a ceiling of $8. The community's schools have been converted to the open-classroom system, which Raza Unida activists claim has slashed the dropout rate and substantially increased the number of college-bound graduates.

Party activists have also laid the groundwork for a community-owned farm. "We were going to establish a corporation of the farm-workers themselves," explains David Ojeda of the Zavala County Economic Development Corporation. "We had $50,000 seed money from the CSA and had picked out 1,000 acres for the farm. The project would have established about seventy-five permanent jobs and two hundred seasonal ones, and we had plans to expand from there. But the plan was dependent on more CSA money to get it going, and Briscoe sued to get the funds cut off." A community housing program has also been stalled for lack of funds.

But Crystal City has been somewhat more fortunate with its energy program. "When Judge Gutierrez declared this a disaster area," says Ojeda, "we went to Army surplus stations to look for blankets. Somebody brought back twenty of those wood heaters used in the Korean War. Once the community found out about them, everybody wanted one."

But the system is not without its problems. Though the stoves provided adequate heat in the winter, they tend to overheat a house when used for cooking during the scorching south Texas summer, and must usually be moved outside. And many families live in HUD homes, where wood stoves are officially banned. Though the stoves are generally popular, other families have resisted them, preferring more "modern" methods. Gutierrez, in discussing long-range solutions, expressed a preference for town-owned gas and oil wells.

Nonetheless, in the absence of capital to undertake drilling, Crystal seems likely to deepen its commitment to wood energy. Local energy experts claim mesquite grows rapidly and exists in such abundance as to be a plague to ranchers. Pliny Fisk, an Austin-based alternative energy pioneer, has designed an inexpensive solar water-heating system which can be hooked into the wood stoves, and which is now operating on one Crystal house—with more to come.

If the pattern set by La Raza Unida in Crystal City is any indication, various community solutions to the energy crisis may be forthcoming— and they may involve much more than just the price of gas.

☆☆☆

Across the state, opponents of the South Texas Nuclear Project were trying to pull Austin's municipal utility out of a 16 percent share in the plant through a series of referendums. The final, crucial ballot came just after Three Mile Island, and failed by fewer than 1000 votes out of some 51,000.

But tragedy stalked the campaign. A series of threats and violence plagued antinuclear organizers. In mid-April, one was murdered:

Murder of Texas Nuclear Foe Stirs Suspicions
(Pacific News Service, April 30, 1979)

A DOUBLE SHOOTING IN HOUSTON that resulted in the death of anti-nuclear activist Michael Eakin has sent waves of alarm through the Texas antinuclear community.

The April 14 murder-assault came amidst a wave of threats and violence aimed at local antinuclear organizers.

Eakin, twenty-eight, was former editor of the University of Texas *Texan*, the nation's biggest college daily. He was also founder of the Austin *Sun*, a local weekly, and an active organizer against atomic power.

The shooting took place on a side street in Houston's Montrose district at 11:30 P.M. as Eakin and Dila Davis, forty-two, returned to Eakin's Mustang from a show at the nearby Texas Opry House. According to Davis, who is the only known witness to the shooting, no one else was walking on the street.

As Eakin and Davis entered the car, one or more assailants pumped four to six bullets through the driver's side window. Eakin was hit in the arm, chest, and throat; Davis was hit in the jaw.

"They must have been waiting for us," Davis said in a telephone interview. There were, she said, "flashes of light and breaking glass, and Michael yelled at me to duck down. I was hit and slid down. Michael got out and tried to run around to the other side to help me. But he never made it."

According to Davis, it took "an eternity" until someone from the neighborhood summoned help. She and Eakin were then taken to separate hospitals. Eakin died several hours later in surgery. Davis, who is the mother of two grown children, was hospitalized with a bullet lodged between her esophagus and backbone, about an eighth of an inch from her spine. Doctors decided an operation would be too risky, and allowed her to return home with the bullet still lodged in her body.

Both Eakin and Davis were well-known in the state-wide campaign against reactors being built at Glen Rose, near Fort Worth, and at Matagorda, outside Bay City. The attack on them came at the crest of a wave of beatings, tire slashings, car trashings, threatening phone calls, and house ransackings that have plagued three Texas cities since January. According to Jeff Jones, a long-time local activist and former president of the University of Texas student body, "There's no doubt that there's an organized campaign going on here to scare off the antinuclear movement. The atmosphere is heavier than it ever was during Vietnam."

Eakin had been active in the antiwar movement and had made a reputation as an investigative reporter, editorializing for public transit and against a controversial highway that would have bisected an Austin park. He also did some of the first reporting on CIA involvement in the Chilean coup against Salvadore Allende. "Michael was one of these people who always had his eye on the future," says Mary Walsh, a long-time friend of Eakin's who now works for the White House. "That's why he got so interested in nuclear power."

According to Jim Hightower, editor of the *Texas Observer*, Eakin was working on a feature about Mexican oil at the time of his death. Other sources indicate he may also have been following up a rash of stories about faulty inspection procedures at the South Texas Nuclear Project.

The 2500 -megawatt Westinghouse plant, scheduled to open in 1981, is being built by Brown and Root. The project has been plagued with reports of intimidation and violence at the site. On April 5, San Antonio Congressman Henry Gonzalez wrote U.S. Attorney General Griffen Bell a formal complaint, demanding an investigation: "I have received reports that inspectors at the South Texas Nuclear Project have been subject to various kinds of harassment and intimidation," wrote Gonzalez. "One construction employee, a concrete foreman by the name of Hinds, was killed in a strange, apparent burglary attempt at a beach house not far from the site. The person who shot him was "no-billed," it being alleged that Hinds was in the act of committing a burglary. An inspector named Perry claims to have been fired for insisting that safety standards in his section be met. Another, named Swayze, claims in a lawsuit that he was fired for overzealous inspections. Other less specific allegations have been reported to me, indicating that inspection documents have been falsified and that inspector initials have been forged."

The South Texas Project has been seriously threatened by the Austin environmental community, of which Eakin was an active member. On April 7 they narrowly missed ending Austin's 16 percent ownership of the project, losing a referendum by 936 votes out of more than 51,000.

Had Austin pulled out, San Antonio, which holds a 24 percent share, might well have followed, leaving the $2 billion project in serious trouble.

But by that time the pronuclear campaign of intimidation was already underway. On January 22, Tod Samusson, a member of Austin Citizens for Economic Energy, and a friend of Eakin's, was beaten in a parking lot outside an Austin convenience store by a man who had been peeling a "No Nukes" bumper sticker off Samusson's car.

Samusson was attacked from behind on a crowded street that following February 14, then punched out in a parking lot on the evening of March 22. Three days later his house was broken into and he began receiving a series of phone calls threatening him for his antinuclear work. On April 9 he was jumped by two assailants in an alleyway outside the office of the Texas Mobilization for Survival. By this time Austin solar activists could document nearly a score of threats or violent attacks, and organizers in Dallas and Fort Worth reported similar incidents.

But Eakin's was the first shooting. "Michael was a perfect target," says a close friend. "He was isolated in Houston, and whoever shot him would know that the Austin community would have a hard time getting Harris County officials to investigate."

Dila Davis may also have been a prime target. Davis is a long-time employee of the Applied Research Laboratory, which shares facilities with the Balcones Research Center and the only radioactive waste dump in Austin. The dump contains residues from research done for the space program, and from a small reactor on the University of Texas campus. Two demonstrations have been held at Balcones, one several days before the shooting. Davis was active in organizing both of them, though she did not speak.

Houston police have no leads. Random violence is common in Texas. "It's been suggested it [the shooting] might have something to do with the antinuclear work," says Detective Johnny Bonds. "We're not ruling that out. It's possible, not probable. We have a large murder rate in Houston. Last year we had 450 murders. At the rate we're going now, we're going to beat that this year."

As in the 1974 death of plutonium worker Karen Silkwood, there is as yet no direct evidence linking the Eakin murder to pronuclear forces, and observers in general discount the possibility.

But Texas nuclear opponents feel the circumstantial evidence, the lack of any other possible motive (no robbery accompanied the shooting), and the context in which the shooting took place are overwhelming. "We've been waging a very successful campaign down here," says Tod Samusson. "And I think the industry has been threatened by us. It's my opinion that the shooting was definitely related to nuclear power.

"We are not going to lay low," Samusson added. "This is making people tighter and more united. A lot of people are going to keep at this nuclear thing until it's gone."

7
Nuclear Exports: How Do You Say "Three Mile Island" in Tagalog?

☆☆☆

The United States has not limited itself to building reactors within its own borders. From the very earliest days of atomic development, exports have been a major part of the business. Pressure to send plants abroad has become particularly strong in recent years, because of the virtual shutdown of domestic orders.

The overseas momentum has also been given a boost by some of America's most imaginative salespeople. Dr. Edward Teller, for example, has toured Southeast Asia at least twice, selling reactors. Among other things, he told the oil-rich Indonesians they should trade all their fossil fuel resources for atomic reactors. He also told a Malaysian audience they might consider building a plant underwater. According to the January 11, 1973, issue of the *Straits Times* of Singapore, Teller told an audience at the University of Malaysia, "I was once chairman of a committee for reactor [accident] prevention and found that the problem of reactor malfunction is almost zero. We have succeeded so far in preventing accidents from happening, and they must never happen because the danger is infinite.''

Despite that infinite danger, Teller felt that the public fear of a nuclear malfunction had been "exaggerated." But if people were still not convinced of the "relative nondanger of a nuclear reactor," he added, then the plant could be built underground or underwater. "Underground," he explained, "a nuclear malfunction will be trapped, and underwater, dilution of nuclear material takes place to make it more or less harmless."

Not so harmless was the 1974 explosion by the Indian government of a nuclear "device" made from wastes produced by its Tarapur reactor. Soon thereafter pathfinding journalist Paul Jacobs documented deadly radioactive spills from that reactor into the Indian Ocean. He also discovered records indicating that workers at the plant were handling radioactive wastes with bamboo poles. The workers themselves were on the job dressed in loincloths, and many of them regularly received heavy doses of radiation.

In the summer of 1976 Amy Wainer, director of the Institute for Appropriate Health Care in Massachusetts, and I tried to interview a high official at the Atomic Energy Department of India. At the time Indira Gandhi was both prime minister and director of India's atomic program. The official threatened to have us arrested for even inquiring about India's nuclear development. Then, when we photographed the department's office building, we were again threatened with arrest. All buildings related to the atomic program, we were told, were "military installations."

Fortunately, rising costs, general political instability, and second thoughts by many reactor producers have produced a significant downturn in overseas orders. In the spring of 1979, Thailand canceled its plans to go nuclear. But the Philippines persisted, and their case began to look like a pivotal case:

Radioactive Roulette
(*Mother Jones*, August, 1979)

WOULD YOU LIKE YOUR COUNTRY TO BUY A NUCLEAR POWER PLANT surrounded by active volcanoes, a few miles from a major earthquake fault, in the midst of an area washed by tidal waves? Nicky Perlas wouldn't. Would you like the United States government to help sell another country one? That's what Nicky Perlas wants to prevent. Perlas is a representative of the Philippine Movement for Environmental Protection, and his residence in the Philippines is about sixty-five miles from the proposed reactor. He has been working in the United States for several years now, trying to keep the Carter administration from allowing the sale of the 620-megawatt, $1.1 billion Westinghouse reactor to the Philippine dictatorship of Ferdinand Marcos.

The Nuclear Regulatory Commission is currently engaged in deciding whether to give Westinghouse a license to export the reactor, and whether to allow the key components to be shipped abroad in the interim. Approval of the Philippines deal has been held up for months in a storm of controversy over political, environmental, and health and safety issues. And in March, construction on the site was at least temporarily stopped.

In May, the staff of the Nuclear Regulatory Commission recommended that the Commission grant Westinghouse interim approval to ship sensitive components abroad. The five-member commission is expected to act on the recommendation by September. At around the same time the Carter administration will decide whether or not to demand a formal environmental impact statement on the plant. The stakes are extremely high, because the decision will affect the entire U.S. pattern of selling nuclear power plants to "LDC's"—Less Developed Countries, in the current Washington jargon.

In the past, judgements about the environmental impact of reactors have been viewed as the sole domain of the purchasing country. But a recent Carter administration order extends the U.S. environmental review process to major federal actions abroad. Westinghouse has been accused of paying off officials close to Philippine President Ferdinand Marcos. In addition, dangers at the site pose a grave threat to Filipinos as well as Americans at nearby U.S. military installations. The Westinghouse application is the first nuclear export application to fall under the new environmental review requirements, and it will test the limits of U.S. efforts to monitor the terms of nuclear sales abroad.

There is strong opposition to the plant in the Philippines and in the United States. The NRC hearings were delayed for eight months, partly because of the receipt of 200 petitions from environmental groups and to the impact of "an emotional outpouring of letters from hundreds of Filipino children," according to an agency official quoted in the *Washington Star*; and there is evidence of violent suppression of dissent by the Marcos dictatorship.

For his opposition to the plant Perlas has been branded a "leftist" by Congressman Larry McDonald (a John Birch Democrat from Georgia), a charge that made front page news in Manila, Perlas's home town. Charges that the Marcos regime may have murdered Ernesto Nazareno, an antinuclear organizer in the Philippines, makes Perlas anxious over the future of those who may try to protest the deal. "In view of what Carter says about human rights," Perlas declares, "we are asking him to stop this sale."

The fight over the Philippine deal underscores the already heated controversy over the international reactor trade in general. This *Mother Jones* investigation casts real doubts on the motivation of the dictatorial regimes which seek to buy nuclear power plants, as well as the motivations of those in the United States who profit from the sales.

As the nuclear industry has encountered more difficulties in selling reactors to disgruntled American communities, it has turned to sales in countries where dissent can be silenced and political payoffs can bring quick results.

The proliferation of these "runaway reactors" abroad is part of an economic pattern which leads to the loss of American jobs, while reinforcing the cycle of poverty in Third World nations. The spectre of spreading radioactive contamination and access to atomic weaponry makes this the most dangerous trade in the world today.

In the course of our overseas research, *Mother Jones* has learned that at least one American-made reactor, at Bang Ken, in Thailand, may have already leaked significant amounts of radiation into the atmosphere and water supply of the city of Bangkok. It has also learned from a report, never published in the United States, that in the mid-1970s the Thai government deemed atomic development as being costly, potentially hazardous, environmentally questionable, and an unwelcome extension of foreign domination into its internal affairs. In late 1978 the Thais announced that because of "technical problems" they would stick with native fossil fuels instead. Such a decision might well be the rule rather than the exception in LDCs around the world if their governments were free of dictatorial regimes more interested in bombs than food.

As of now the Philippine reactor deal is scheduled to be financed by the biggest Export-Import Bank loan in U.S. history. But the Philippines is already $4.5 billion in debt to foreign lenders. Without this loan the reactor purchase would be impossible for them. Since soon after the project began in the early seventies, the Bataan deal has been haunted by tales of corruption on the part of Philippine officials and by demands that government investments in the Philippines go for agriculture, not electrical production.

Tragically, the Philippine story seems no exception to the pattern of nuclear exports. Most underdeveloped countries do need food more than electricity. And in terms of human rights, corruption, and the potential military uses of atomic power, "the list of developing countries reads like a 'Who's Who' from Amnesty International," according to Keiki Kehoe of the Center for Development Policy in Washington. "Brazil, Argentina, South Korea, South Africa—they all seem most concerned about prestige and bombs."

Atoms in Pieces

As of late 1978, the prospective export market for American reactors stood at $36 billion, representing the sale of 59 large power plants abroad. That's just 13 less than the total number operating in the U.S. today.

The domestic industry is in the midst of a drastic downturn. Sales of reactors to U.S. utilities have plunged from 117 ordered between 1970 and 1975 to just 11 from 1975 to 1978. Cancellations and postponements in 1976 and 1977 alone came to 51 power plants.

The industry as a whole entered 1979 in the mood for a counterattack, desperately trying to revive itself.

But then came Three Mile Island. In the wake of the accident, with its obvious chilling effect, reactor producers may be hoping for some help from overseas. But even that angle has been less than promising. In 1978 only two LDC orders came in, from South Korea.

The fall of Iran's Shah, who had planned to build three dozen nuclear plants, many of them American-made, hasn't helped. Should the Philippine deal fall through, the hopes of American producers for salvation through the export market could all but die.

How Did We Get Here?

The birth of the so-called peaceful atom was an international event. It came in the form of Dwight Eisenhower's "Atoms for Peace" speech to the United Nations in 1953. Eisenhower promised the world that the force which obliterated Hiroshima and Nagasaki would make rich nations richer, and provide poor ones with cheap, infinite energy for industrialization. The U.S. helped distribute small test reactors around the globe, hoping to pave the way for bigger commercial ones in the years ahead.

Even at the beginning, the idea of exporting reactors was not all that altruistic. It was predicated on the notion that everyone on the planet wanted the kind of high-tech life-style Americans could introduce them to. Urbanization, political centralization, and smog were the eventual results in the Third World nations that did industrialize under U.S. tutelage.

American reactor producers got their start through the export trade. The first five reactors Westinghouse sold went, with the help of the Export-Import Bank, to foreign buyers. Westinghouse and General Electric, the two biggest manufacturers, soon established strongholds in Canada, France, West Germany, Sweden, and Japan by licensing multinational companies there to act as go-betweens for the sale of what was essentially American technology. Exporting reactors thus not only gave the American industry new markets, it gave atomic power a new credibility in the U.S. and abroad.

Selling nuclear power plants abroad, however, turned out to be a little different from selling shoes. All commercial reactors produce plutonium wastes, which can be made into nuclear bombs. Until the mid-seventies, the U.S. believed Third World nations would probably never—or at least not soon—obtain the technology for doing it.

But in 1975 India exploded a nuclear device made from waste plutonium produced in its reactor at Tarapur. Two years later, Pakistan's imprisoned former premier, Ali Bhutto, confirmed that his intent in trying to buy a French reprocessing plant was to build an atomic bomb. Indeed, most LDCs that do buy nuclear plants *are* looking for prestige, the prestige of joining the no-longer exclusive nuclear club. That was one of the observations in the 1975 Barber Report, a government-sponsored study that was highly critical of LDC nuclear development. The Barber Report also estimated that by 1990, forty-six nations could have collective access to some 15,000 kilograms of plutonium yearly—enough to make 3,000 Nagasaki-size bombs, year after year.

The potential diversion of plutonium or the covert enrichment of uranium for use in illicit weapons is supposed to be monitored throughout the world by the International Atomic Energy Agency (IAEA). The IAEA was set up to do this in 1953 when the Atoms for Peace program was introduced. But it's a paper tiger. Notoriously ill-funded, the Agency has fewer that a hundred inspectors. Morris Rosen, of the IAEA, describes his own agency's efficacy in LDCs as "subminimal."

In 1976 some fifty pounds of fissionable material disappeared from a Taiwan reactor, enough to make five bombs. The Taiwanese government says the material has been moved to "another site." IAEA has been unable to retrieve it. "I don't think they [the IAEA] can do an adequate job," says Dr. Ichikawa Sadao, a noted geneticist in the Japanese antinuclear movement. "There just aren't enough resources."

Nuclear Salesmanship

Philippines dictator Ferdinand Marcos was talking of a major nuclear complex at Bataan, the site of a famous World War II battle, as early as 1968. By the early seventies negotiations with the U.S. Export-Import Bank and the State Department had begun in earnest. But already rumors of corruption were seeping into the negotiations. And the price was skyrocketing.

In June, 1974, Westinghouse had presented Marcos with plans for twin 600-megawatt reactors at a total cost of $500 million. By June, 1975, the price had more than tripled, to $1.6 billion for the two units. Current estimates stand at $1.1 billion for *one* unit, four times the original estimate.

The Philippine National Computer Center later compared the proposed Westinghouse plant with similar projects and concluded it was overpriced by at least $75 million. An official of the Philippine

Nuclear Power Corporation then admitted that "the decision to choose Westinghouse was a political decision" made within the Philippines.

Westinghouse had hired a business agent named Herminio Disini to oversee its bidding. Disini is a frequent golfing partner of Marcos's. His wife is a cousin and personal physician to Marcos's wife, Imelda, the Governor of Manila. Disini is flamboyant figure on the Manila scene and his rise to millionaire status has been meteoric, climbing from obscurity to become head of the Herdis Conglomerate, an umbrella for some thirty Filipino companies.

Just prior to public announcement of the reactor deal, Disini purchased the other company acting as agent for Westinghouse in Manila, Asia Industries, from U.S. Industries of New York. With the sale official, Disini's commission became an object of wide speculation. Westinghouse refuses to divulge the actual figures, but bankers' estimates of Disini's profit on the arrangement run between $4 million and $35 million.

Disini's Power Contractors Corporation became principal subcontractor for the project. Disini's Summa Insurance Corporation was chosen to handle the builders' risk insurance, a policy involving $693 million, possibly the biggest contract of its kind in Asia. That deal raised some eyebrows because Summa is both new and thin on assets. Other Herdis subsidiaries also landed contracts for communications and construction work at the site.

In early 1978 the web of Herdis' nuclear connections found its way into the world press along with photos of Marcos and Disini smiling together at Manila's fashionable Wack Wack Golf Club (of which Disini is president).

In a grandiose public gesture, Marcos ordered an investigation into the Westinghouse deal, then divested Herdis of three of its companies, all of them of marginal value. The U.S. Securities and Exchange Commission and the Justice Department have been conducting investigations of their own. Incensed by the scandals surrounding the project, Congressman Clarence Long (D-Baltimore) held hearings on the Philippine reactor in early 1978. These led to Long's proposal that the Export-Import Bank be banned from financing nuclear sales.

Exim's burden in the Philippines sale was to be a staggering, unprecedented $644 million in loans and guarantees. Yet there were strong indications that nuclear power was not the most economical way to generate electricity there. A major 1976 study by the Philippines Power Company, the national utility, showed that nuclear energy would cost them $1500 per kilowatt. Hydropower came in at $1000, coal at $800, and geothermal, which is undergoing rapid advances in the Philippines, at $900.

With the cost of the Philippine reactor still escalating, nuclear power looks like a bad bargain. But as Congressman Long complained in an interview, "The more it costs, the more they want it. Shouldn't a country that faces massive unemployment and food shortages be more concerned with developing its labor-intensive agricultural sector?"

Good question. The $1.1 billion price tag currently on the Bataan plant represents three times the annual investment in agriculture in a nation where 70 percent of the population is rural, 57 percent of the households lack running water, and eight million children suffer from malnutrition. Yet the biggest single capital investment in the country's history is going for a giant generator meant to produce a very specialized form of energy directed, it seems, not toward the needs of the local population, but to the service of the nation selling the plant. "The reactor is not designed to supply electricity to our people," claims Nicky Perlas. "It's for Clark Air Force Base and the Subic Naval Base and the Bataan free trade zone, where foreign companies make textiles for foreign markets—most of them American."

Environmental Impact

So far the Nuclear Regulatory Commission has refused to complete the deal. The major sticking point is whether the Carter Administration, or the courts, will require an environmental impact statement on the plant. Ecological problems with the project are mindboggling:

• Local farmers complain that massive silt runoffs from the site have already wrecked rice fields and fish farms where "bangus" fingerlings provide an important protein source.
• The twenty families pushed off the three hundred-acre site to make way for the bulldozers have long since relocated, but well over ten thousand people still remain in the near environs. Barrio Nagbalayong, a town of three thousand, is just four kilometers away. An accident at Bataan could do to it what Three Mile Island threatened to do to Middletown, and more.

• In August, 1968, the immediate area experienced an earthquake. In October, 1971, a tidal wave washed close to the site. Five volcanoes, four of them active, sit within a hundred miles of the site, the nearest one just ten miles away. According to Friends of the Earth, ash falling from an eruption at any one of them "could cover the site like heavy snowballs, clogging filters and cooling ponds and raising havoc with the nuclear plant's complex machinery."

• Philippine Atomic Energy Commissioner Librado Ibe has assured the public that "the construction of the nuclear power plant will be pushed through even if no storage site is found for the plant's radioactive wastes." One national utility executive has suggested that, if nothing else, nuclear wastes could simply be dumped into the ocean surrounding the country.

Critics of the Bataan reactor have not fared well. According to Friends of the Filipino People, local farmers who oppose the project are regularly terrorized, their homes broken into and their families threatened. One Protestant minister who spoke out at a public meeting was threatened with arrest and then transferred by his church. This winter a thousand soldiers and police raided the nearby town of Morong, breaking into houses in search, they said, of antinuclear activists.

The most serious of these human rights violations concerns the disappearance of Ernesto Nazareno. Nazareno was a twenty-three-year-old construction worker who was accused of trying to organize laborers on the Bataan site. Nazareno was arrested on March 13, 1978, and then tortured. He was released on May 3 and ordered to report back to military officials periodically. He did so the following June 14, and has never been seen again.

Thailand Says No

Because of its unique political history, Thailand has also been a hotbed of nuclear controversy.

The original site of Thailand's only proposed commercial-sized reactor plant can be found about an hour's journey south of Bangkok, where water buffalo wander the dusty streets of a tiny farming village called Ban Ao Phrai. A mile away, guarded by a soldier with a submachine gun, stands a four-sided weather tower designed to test wind direction for the proposed plant.

In 1976, when we visited the site, Thailand was in a unique political situation. Three years earlier the vicious military regime of Thanom Kittikachorn had been overthrown, opening the way for a liberal constitutional democracy under the prime ministership of a university professor named Sanya. When Sanya came to power he commissioned a study on the Ban Ao Phai project. The previous military regime had strongly favored it. But a university group came back with some harsh conclusions: the reactor could barely compete with coal or oil; there would be no long-term guarantee of uranium supply; no guarantee that environmental problems (like hot water emissions into the Gulf of Siam) would be solved or that Thailand could train a sufficient number of technicians to run the plant. The final report advised that the project be postponed, and Ban Ao Phai was put on the shelf.

The project stayed on the shelf until the winter of 1978–79, when the Thai government, citing the economic drawbacks of atomic power, virtually canceled the project by deciding to produce its new electrical capacity by conventional means.

The Sanya Report that led to that decision remains a seminal Third World renunciation of nuclear development. A similar report from a panel in any of the other LDCs now considering reactors might well result in a long string of stories like that of Ernesto Nazareno.

But Thailand, at the time of the Sanya report, already had an operating reactor, a small test facility on the campus of Kasetsart University, north of Bangkok. Built by Curtiss-Wright, an American company, the 1-megawatt Bang Ken facility was designed to produce medical isotopes and train local technicians. (In 1977 the reactor was upgraded to 2 megawatts by the General Atomic Company of San Diego.)

During the Sanya era, several Kasetsart graduate students gained access to the plant's operating files and issued a pamphlet titled, "Top Secret: The Leaking of the Atomic Reactor at Bang Ken." The document created a national uproar. According to the pamphlet, data from Atomic Energy for Peace (AEP, the plant's operator) indicated that the reactor had been emitting 10 to 600 times the radiation level considered "normal."

The students said the AEP documents confirmed regular dumpings of radioactive materials into *klangs* (canals) which were used to irrigate rice paddies and which also drained into the Chao Phrya River, which flowed through Bangkok itself, and is its chief source of drinking water.

After a friend translated the pamphlet for us while we were in Thailand in 1976, I called a high official in the Thai Energy Generating Authority. He was indeed familiar with it. But even in a period of constitutional democracy, discussion of the nuclear issue was none too welcome. The students "couldn't read the documents they were citing," he snapped.

I then asked about a passage in the Sanya Report that said Thailand's Bang Ken had been "leaking for two years before it was repaired . . .nor was the populace warned of the possible danger. The agency involved kept quiet about it."

The official demanded to know how I got the documents, who translated them for me and whose phone I was using.

I refused to tell him.

He refused to comment further.

Optimistic Watchdogs

I first met Richard Kennedy, a stocky, six-foot Nuclear Regulatory Commissioner, at NRC headquarters in late June of 1978. Outside, 350 demonstrators were camped on the sidewalk protesting construction of the Seabrook reactors. Richard Kennedy saw the bright side of the situation. "Reminds me of seven years of my life, with Dr. Kissinger on the National Security Council," Kennedy said, looking somewhat wistfully through the plate glass at the angry crowd.

At the time we talked, Kennedy had just voted to allow the sale of American uranium fuel to India. India had obviously violated international guidelines by using reactor-produced materials to build its atomic weapon. Opponents of the sale argued that the United States was legally bound to stop selling India nuclear fuel.

Commissioners Kennedy and Joseph Hendrie disagreed, insisting that to stop selling the fuel would only encourage India to buy it elsewhere, violating prior U.S.-India safeguard agreements. India would then be free to use for bombs *all* the materials the U.S. had previously sold them. The Carter administration agreed, and the sale went through, underscoring a classic nuclear "Catch-22." Once you start selling, how do you stop?

I put that question to Griff Ellison a few blocks away, at the elephantine headquarters of the U.S. Export-Import Bank. Exim is financing a lot more than the Philippine reactor; in fact, every reactor the U.S. has exported has been Exim-financed. Griff Ellison doesn't sign the checks, of course, he just explains why they were written. He is a public relations man for Exim, a soft-spoken Virginian in his early thirties, once an editor of *The Progressive*. I remember him as the co-founder of Vietnam Veterans Against the War. It's no secret Carter filled numerous Washington offices with veterans of the antiwar movement. Ellison and I shared some mutual friends and experiences.

But times change. As we approached the subject of nuclear power, our conversation became uneasy. "I've traveled around a lot and I've thought about it a great deal," he said. "It's a question of short-term energy options. What else do they have? I don't think we have the right to deny developing nations access to nuclear technology. It's a choice they should be allowed to make for themselves."

As for Exim's role, Ellison emphasized it was merely a bank, following the winds of the world market. "There are restrictions on loans for reprocessing and enrichment," he explained. "But our chief function is to promote foreign trade, and nuclear power plants are a part of that. If the U.S. doesn't sell them, the French or Germans will. The competition is fierce."

The spectre of fierce competition is almost as common in nuclear industry rhetoric as repudiations of solar power. Britain, France, West Germany, Russia, Japan, and Canada are all, in one form or another, in the nuclear export business. The French and Germans in particular have won LDC orders in recent years that Westinghouse and G.E. thought should have been theirs. But the notion of "fierce competition" is a little misleading.

"We gave the French and Germans this technology," says Robert Alvarez of the Environmental Policy Center, who has examined the ownership of overseas nuclear firms. "We set them up, and Westinghouse and G.E. still profit from their business. For American producers now to use their competition as an excuse to push more reactors isn't quite honest."

According to Alvarez, the French firm Sogerca is totally owned by General Electric. Brown-Babcock-Bovari is largely owned by Babcock and Wilcox. U.S. interests are also strong in Mitsubishi and Hitachi, two Japanese exporters who operate on license from Westinghouse and G.E respectively. Many components of the Canadian CANDU reactor are made in the U.S.

American influence on the overseas producers, however, has definitely slipped in recent years. And both the British and Soviet export efforts have always been free of American ownership. Thus there is more than a grain of truth in it when American reactor producers complain about independent competition abroad. The Germans, French, British, Soviets, and Canadians have all been happy to line up against American producers in the race for LDC atomic billions.

Indeed, says Gloria Duffy, who did a major survey of the world trade for the Council on Economic Priorities, "The French and Germans are very eager to sell the reactors. They seem less concerned that the plants will lead to weapons proliferation, and the European antinuclear movements don't seem as strong on the export issue. The economies there are more dependent on exports, and it's easier for European leaders to make the argument that 'exports are our life-blood.' "

How much sustenance they can actually garner from the export trade is another question. Germany's $8 billion atomic deal with Brazil had fallen under a cloud as the Brazilian economy has been faltering, and the Germans' ability to deliver promised reprocessing technology now seems doubtful. The joke in international circles goes that Germany sold technology it couldn't produce for money the Brazilians didn't have. Canada experienced embarassment and pain when deals with Korea and Argentina yielded multimillion-dollar commissions scandals.

Yet exporters continue to fight over the LDC sales, and the competition is a far cry from Adam Smith's free market.

But U.S. producers still vastly outstrip all others in their ability to produce large quantities of reactors. With the Soviet Union, the Americans hold a virtual monopoly on enriched uranium. The lending power of the Export-Import bank to promote foreign buying from American firms remains prodigious, as does the world-wide power of the American government. Many of the purchasing regimes would fall without American backing. Says Gloria Duffy: "The French have now canceled their sale to Pakistan, apparently because of U.S. pressure, and South Korea pulled out of a deal for the same reason, all of which shoots industry arguments about the decline of U.S. influence."

Also off the mark are industry arguments that reactor sales abroad create jobs at home. In fact, the opposite can often prove true. Cheap industrial exports can only be produced in foreign countries with subsidized electrical power. "The idea is to provide centralized electricity for subsidized trade zones like Bataan,'' says Keiki Kehoe of the Center for Development Policy, ''American capital will pour in to build the factories which will use nuclear electricity to manufacture goods for export. The only role for the local population is cheap labor—in direct competition with American workers. Meanwhile, the nuke demands repayment of a huge capital debt and forces the LDCs off their agricultural base.''

Thus reactor exports don't preserve U.S. jobs; rather, they guarantee runaway shops.

Magnifying Questions

The export of nuclear power presents a picture which offers little financial or military security. Most LDC purchasers have been relatively large nations with strong regional rivalries—Argentina and Brazil, Iraq and Iran, Korea and Taiwan, India and Pakistan, South Africa. Unpleasant surprises are an inevitability.

In 1975, as U.S. troops beat a wild retreat from Saigon, one of their last frantic chores was to retrieve fissionable material from the small American-made test reactor at Da Lat. The plant has been dormant since. But India is now training Vietnamese technicians to refire it. China, on Vietnam's border, has the atomic bomb now. Vietnam does not . . . yet.

The United States has sold Taiwan a half-dozen commercial reactors. Since the U.S. has recognized the People's Republic—and derecognized Taiwan—what happens to those reactors now? Are the original nuclear safeguard agreements still in effect? What will prevent Taiwan from using those commercial reactors to make nuclear bombs?

And what happens when, inevitably, human error and faulty construction lead to a Three Mile Island at an LDC reactor. Will a crack team of experts be available to fly in, on short notice, with unlimited resources, to save the day? Will the world media be there to guarantee saturation coverage, and to monitor the evacuation of the neighborhood?

Or will we hear about it only when the radioactive cloud makes its way, as they all do, around the globe?

In a world of corrupt regimes and political instability there are often no easy answers. The atom simply has a way of magnifying the questions.

8
Japan's Nuclear Crisis— And Narita

☆☆☆

Japan is the world's number two generator of commercial atomic power. In mid-1979 it had nineteen operating reactors, all but one of which were American-made. Fourteen more were under construction or in the advanced planning stage.

At the time of the Three Mile Island incident, the antinuclear campaign in Japan was already well under way. The list of the archipelago's environmental problems, however, was a crowded one, with catastrophes ranging from mercury, cadmium, and chromium poisoning to pollution-caused asthma to people keeling over unconscious in Tokyo's summer smog.

But the official push for atomic power was particularly forceful. With no domestic fossil fuels to speak of, Japan's industrial elite saw nuclear energy as the hope of the future. Its development, said a government white paper in 1973, "will probably be our greatest gift to our grandchildren." Plans were announced to construct by 1985, 60,000 megawatts of atomic capacity— representing fully 25 percent of the nation's total energy consumption.

But political opposition cut the target in half. In 1979, in the wake of Three Mile Island, even that much atomic energy began to seem unrealistic, as public outcry reached fever pitch. Among other things, Takeshi Hirayama, chief epidemiologist at the Japan National Cancer Center, warned that "the popularization of nuclear reactors will lead to such an increase in radioactivity in the environment that it will cause approximately ten thousand more cancer deaths a year by the mid-1980s." Prominent scientists also warned against the impact of hot water emissions on the nation's vital fishing industry. "Judging by the present furor," wrote correspondent Henri Hymans from Tokyo in May, 1979, the government "will be hard put to come anywhere near its proclaimed targets."

Japan's Nuclear Crisis
(*The Progressive*, November, 1976)
&
No Nukes in Japan
(*WIN* Magazine, April, 1976)

THE FIRST JAPANESE REACTOR opened at Tokai in 1966. A British gas-cooled model, it remains the only major plant of non-U.S. design in the archipelago.

Despite the national "nuclear allergy" contracted at Hiroshima and Nagasaki, the Japanese people apparently welcomed the "peaceful atom." According to the *Japan Times*, Tokai was greeted as "a symbol of the peace-loving nation that renounced war under constitution." In the midst of unprecedented industrial expansion, communities throughout Japan competed for nukes and the tax and job benefits they were told to expect. The story is much the same as in the U.S. where people in small towns have often welcomed nukes with open arms.

In March, 1970, the first full-scale plant opened at Tsuruga. A second opened within the year at Mihama. In 1971 the government announced plans for sixteen more plants, and added three more to the list in 1972. By 1976 there were a dozen operable plants in Japan, with the government talking in terms of fifty by the year 2000. All the plants except Tokai were designed by General Electric or Westinghouse.

With an annual electric consumption growth rate well over 12 percent, and projections for an over-all economic growth rate of 7 percent, Japan's nine privately owned utilities were totally sold on the need for nukes. Japan has no coal or oil to speak of. With strong assurances from U.S. reactor producers, the Japanese governmental-industrial complex committed itself to being 25 percent dependent on nuclear electricity by 1985.

In the public's mind, however, the romance with the atom had already been tarnished.

Radioactive Sea

In 1971, fishermen working the seas around Tsuruga began catching abnormally large trepangs (sea cucumbers). Cobalt-60 was subsequently found in the sea water, in mussels, and in pearl oysters near the plant's coolant outlets. Later it was also found in seaweed.

Area residents were greatly alarmed. Japanese eat five to six times as much seafood as Westerners; they rely on it for 50 percent of their protein. There could be no more basic threat to Japanese life than a radioactive sea.

The bad news began to pile up. In March, 1971, radioactive gas was detected leaking from the Tsuruga stacks. In June there was an iodine-131 leak, prompting the removal of fourteen rods.

In June, 1973, highly radioactive waste water leaked from a storage room at Fukushima I. Someone failed to shut a drainage valve, and nearly three cubic meters of "hot" water were misrouted.

In July, 1974, Kansai Electric asked Westinghouse to replace the steam generator of one of Kansai's two Mihama reactors. Mihama I had experienced four major shutdowns in less than four years.

In September, 1974, twenty-one of fifty-five U.S. nukes were closed for an emergency safety check, prompted by a leaky bypass pipe at Illinois' Dresden I plant. Japanese authorities decided to inspect their six boiling water reactors because of their similarity in design and manufacture to the U.S. plant. They found alarmingly similar defects at Tokyo Electric's Fukushima I and at Hamaoka.

In January, 1975, twenty-three U.S. plants shut down for a second emergency check, this time because of an emergency core cooling system problem—again at Dresden. The Japanese checked their

BWR's and once again found critical problems, this time at Tsuruga and at Fukushima I, which was just about to reopen.

At the same time, Kansai's Mihama II was shut because of "wall thinning" of tubes in the reactor cooling system—a possible cause of radioactive releases. In May, further inspections at Mihama II revealed a serious problem that required removal of half the 121 fuel assemblies. Called "bowing," or fuel rod bending, the problem greatly angered Kansai officials, who had been assured by Westinghouse that fuel rod technology had been perfected. Although they would publicly deny it, Kansai officials soon asked Westinghouse for a refund on their Mihama I reactor, which (as of late 1978) had remained shut since 1975.

A month prior to the Mihama II bowing discovery, ten workers at the construction site of the Tokai reprocessing plant were exposed to potentially fatal cobalt-60 radiation.

At Takahama, reactor unit II was shut for a month because jellyfish had fouled the coolant intake pipe. The plant is designed to suck in fifty tons of seawater per second, and swarms of jellyfish attracted by warm waste water had clogged the intake pipes at both Takahama I and II. The problem required the removal of more than twenty thousand jellyfish per day.

While the environmental problems at Japanese plants began to take on a nightmarish quality, some severe economic factors also came into play. Assured by U.S. industry that the technology would pay for itself, Japanese utilities in fact found themselves with a herd of white elephants. Twice in 1974, seven of eight operable reactors were shut down simultaneously. At one point, in the fall of 1975, only one out of twelve operable plants was actually generating. From April until September of 1977, six of Japan's fourteen reactors remained inactive. The year's over-all operations produced a dismal capacity factor of under 47 percent. Construction costs are considerably higher in Japan than in the U.S. With its catastrophic operating record, the Japanese reactor program had lost much of its luster.

Mutsu vs. Fishermen

Economic factors aside, no single recent incident has contributed more to Japanese disaffection with nuclear power than the *Mutsu*.

Mutsu is an 8,214-ton surface ship built by the Japan Nuclear Ship Development Agency as a national showpiece at a cost of $133 million. Begun in the late sixties, it was the only Japanese reactor built without British or U.S. guidance, although Westinghouse was called in at the end for an advisory check.

The project was meant to mark Japan's emergence into the age of independent atomic technology. The project was originally welcomed by the local government of Mutsu Bay, a fishing community in the remote northern reaches of Honshu, Japan's main island. Many felt *Mutsu* would be a boon to the depressed local economy.

But as the ship was being built, local residents began harboring doubts about the unfamiliar nuclear intrusion. By 1972, when *Mutsu* was ready to sail, there was enough public opposition to force Japanese courts to keep the ship anchored in port. Fears that radioactivity would destroy fishing and scallop cultivation brought local people into open, angry protest. Even the government's request to set sail under conventional power and fire the reactor only at sea was not enough.

Finally, in August, 1974, the government had had enough. Moriyama Kinji, head of the Science and Technology Agency, announced arrogantly, "I will never yield to the pressure of demonstrations and red flags. I consider them 'congratulations' on our departure."

But on August 25, three hundred small fishing boats swarmed around the nuclear ship. While twenty thousand people cheered from shore, the fishing people lashed their armada into a giant blockade. When one of them managed to thread a mooring rope through the *Mutsu*'s anchor chain, the ship became an atomic prisoner.

Twelve hours later, Maritime Safety Board patrol boats tried slashing their way through the barricade, only to be beaten back. A tugboat tried hauling the ship out, but the fishermen chopped through the line with hatchets.

Finally, at night, rising typhoon winds forced the blockade to disperse. The *Mutsu* limped out of port, and the fishing people vowed to keep it out.

As the *Mutsu* proceeded to sea and fired its reactor, three "Banzai" cheers came from the fifty-eight man crew. But no sooner did it go critical than radioactivity was detected leaking from the top of the reactor. At 1.4 percent of capacity, the radiation indicator showed 0.2 milliroentgens, a level expected only at 100 percent capacity. At that rate the radiation level would have reached as much as 400,000 times the maximum standard.

Technicians soon surmised that the top reactor shield was inadequate to stop fast neutrons, and plugged it with seventy-five pounds of boron-treated rice balls. Later they filled crew members' socks with polyethylene and covered the reactor with these. Finally, there was no choice but to shut down the reactor. The crew members were now prisoners on a radioactive ship.

Meanwhile, the Mutsu Bay fishing people began laying sandbags at the harbor mouth. The *Mutsu* drifted helplessly for fifty-one days before an agreement was reached. The ship was then allowed to return on condition that a new home port be found within six months. The ship docked safely and was subjected to a thorough examination.

Technicians found the problem "was not a small accident but a serious mistake in the reactor's basic design." A shielding ring at the top of the reactor had been made of metal, but the Westinghouse advisers had recommended that the ring be made of concrete. The Japanese had chosen to ignore the recommendation. The test run proved that the metal shield could not block fast neutron emissions, although it was not clear a concrete shield would either. According to one Japanese critic, "The failure of the test run and later-disclosed serious mistakes in the reactor design indicate unaccomplished nuclear technology in our country, as well as carelessness of the administrative authority to reactor safety."

In 1978, after at least ten towns had turned it down, *Mutsu* was moved to the port city of Sasebo, near Nagasaki, amid massive public protests. The ship was taken there ostensibly for refitting, and to help revive a lagging shipbuilding industry.

The original *Mutsu* incident poked a big hole in Japan's nuclear credibility. A national opinion poll taken before *Mutsu*'s 1974 cruise revealed that 44 percent of the Japanese public thought nuclear power "dangerous"; afterwards, the figure had soared to 77 percent. "The *Mutsu* fiasco really hurt us," admitted a high Tokyo Electric official. "Our years of efforts to persuade local people have been absolutely shot."

One such local area was Kashiwazaki, a farming and fishing community on the Sea of Japan, four hours by train from Tokyo. It sits in the midst of Niigata Prefecture, known as Japan's "rice bowl," famous also for its giant ornamental carp. Its people are peasants of the land and sea who compose the conservative backbone of Japanese society. In September their ancient hillsides turn the thousand shades of a New England fall, while the rice paddies take on a deep, mature green and the carp-raisers take to their mucky ponds to scoop out multicolored fish that can sometimes sell, as collector's items, for up to $30,000 each.

For many years both farming and ocean fishing have been on the decline in Niigata. Most farmers now have part-time or full-time jobs and work their fields in the evening and on holidays. Young people have also been less than eager to risk their lives on the rough Sea of Japan for scant, chancy profits. To make things worse, the dread Minimata disease (mercury poisoning) was found near Niigata City, further discouraging the fishing trade.

So in the late sixties many Kashiwazaki townspeople were understandably happy to find a potential new source of income—a nuclear power plant. In 1968 Tokyo Electric (Todan) announced plans to build at least one reactor, and possibly as many as eight, about five kilometers from Kashiwazaki City, where 40,000 of the town's 80,000 people are clustered. A year later the town readily approved the project, and soon thereafter the company built a sixty-meter weather tower to check wind speed and direction. (Taking no chances, the company put down a very solid four-sided tower with turnbuckles that would be virtually impossible to disengage by hand. The base of the tower is buried deep in concrete).

Construction was scheduled to start in 1972; the plant would open in 1976.

But Tokyo Electric soon ran into unanticipated problems. For one thing, local fishermen began holding out for more money for their fishing rights, which Todan would be required to buy if it was going to dump hot water into the sea. For the old fishermen, it was a last chance to make some money, and they stalled for precious months while the price went up.

Then the area was "infiltrated" by students from Tokyo. Sleeping on floors and in fields, the students brought the first winds of an antinuclear line, and backed up their case with visits from prominent antinuclear scientists.

At first the townspeople resented the students. Among other things, Todan had told them radioactivity might be good for the rice crops and for coloring the carp. There seemed no reason to doubt the company.

But gradually the students' message began to seep in. As the months slipped away, local organizers were made suspicious by the fact that Todan changed the exact plant site five times. The company's geological survey had pronounced the area "stable." But the opposition decided to call in its own experts.

What they found was that the site is in fact a potentially unstable dislocation center that may have been active as recently as 30,000 years ago. The estimate was backed up by an extensive oil drilling survey done in the 1920s. The company had told the public the fault was active 300,000 years ago, and thus it was trapped in an obvious, conscious lie. Suddenly, the plant became an insult.

The scene was further polarized by a national scandal. Niigata Prefecture is the home of former Prime Minister Kekeui Tanaka. Born poor, Tanaka made a fortune as a land dealer and soon became a political hero as well—a local boy who made it big in Tokyo. One reason Todan was so confident about its choice of Kashiwazaki was a heavy banking on Tanaka's popularity there.

But the plan backfired when an investigative reporter found that one of Tanaka's land companies had bought the Kashiwazaki site before the plant was announced, telling locals they were planning a "resort." They then sold the land to Todan for more than twenty-five times what they paid. That and another shady land deal helped Tanaka, a long-time friend of Richard Nixon, out of office.

In 1974 came the *Mutsu*.

By then most of the students had left Kashiwazaki, and the movement was firmly in the hands of the local citizenry. A mid-1975 poll showed 30 percent of Kashiwazaki firmly opposed to the plant, another 50 percent "worried." They have been fighting—successfully—ever since.

☆☆☆

The Kashiwazaki fight is typical of local resistance that has hamstrung atomic development throughout the archipelago. A controversial 1978 court decision polarized the opposition even further. In August, 1973, residents living near the proposed Ikata reactor sued the government in Matsuyama District Court. Over the course of nearly five years the suit involved many of Japan's top scientists.

There was also a dubious shifting of judges at key points in the trial. When the decision came down in the spring of 1978, at least one key judge had heard only a very small portion of the testimony.

So it came as no surprise to nuclear opponents that the verdict favored contruction of the plant. "For many people the Ikata decision meant the end of legal channels," said Dr. Sadao Ichikawa, a geneticist actively involved in the Japanese movement.

The well-publicized Ikata case spread knowledge of the nuclear fight around the country. Three Mile Island polarized things even further. With a growing resistance network focused on communities like Kashiwazaki, organizers were confident that they could turn the Japanese nuclear industry into one big *Mutsu*, under blockade.

Ironically, however, through the early development of the Japanese antinuclear campaign, the key battles were being fought not at the sites, but at an airport forty miles northwest of Tokyo. Though it didn't gain world-wide notoriety until 1978, the confrontation over the New Tokyo International Airport had been disrupting Japan for a decade. As the premier environmental struggle in the archipelago, its ferocity set the style and tone for environmental confrontations throughout Japan. Indeed, the birth of the modern Japanese ecological—and antinuclear—movement came not at a nuclear site, but amidst the brutal, epic battles of the war at "Japan's Vietnam"—Narita:

It's Not an Airport...
The Struggle in Sanrizuka
(*WIN* Magazine, November 11,1976)

Narita, Japan

IN 1965, THE JAPANESE GOVERNMENT decided to build a new international airport.

Like everything else at the time, the decision was heavily tainted by the war in Vietnam. Under a mutual security pact (AMPO), the U.S. uses Japan for a military base. In the sixties, American bombers and troop charters were raining fire on Vietnam, and needed all the runway space they could get. Partly because of American war use, and partly because of Japan's unprecedented economic expansion, the International Airport at Haneda, a few kilometers south of Tokyo, became badly overcrowded.

In 1966, the site for the new airport was chosen—a small farming village called Tomisato, some sixty kilometers from Tokyo Station. But Tomisato has a long history of peasant revolt, and it took only a short period of fierce resistance to convince the government they'd better try somewhere else.

So they chose neighboring Sanrizuka, which had a few things going for it. For one, much of it used to be grazing grounds for the horses of the old imperial lords. About a third of the prospective site still belonged to the Emperor. Hirohito wasn't about to argue with a new airport.

Secondly, Sanrizuka has only been farmed since World War II. For centuries its soft, volcanic soil discouraged farming, and it was only after 1945 that the government could persuade some impoverished peasants to settle there and work the land.

Thus the government reasoned that people who had been on the land so short a time could be easily convinced to move. But what they hadn't counted on was that the peasants had broken their backs for twenty years on that land, working in the daytime for food and returning to their plots to work through the night, usually with borrowed tools. More than once they had been badly misled by government experts who told them to plant the wrong crops at the wrong time. When the government came again with an offer to leave, much of the Sanrizuka peasantry was not impressed.

On June 22, 1966, the government officially announced it had given up on Tomisato and would build the airport at Sanrizuka. Six days later, fifteen hundred Sanrizuka farmers met in a rainstorm and announced their opposition. "We hereby warn the government," they said, "that if they try to force this construction on us they will be driven into a corner and will be forced to dig their own grave."

Two days later, they were joined by farmers from Shibayama, where noise from the jetport was expected to make the fields sound like a Tokyo subway tube. Calling themselves the Hantai Domei

(Opposition League), the farmers spent the first year of resistance under the influence of the Japan Communist Party (JCP), petitioning and conducting peaceful protests and rallies.

Meanwhile Kodan, the private front corporation funded by the government to build the airport, proceeded as if no opposition existed. The airport would open in April, 1971. It would cost 60 billion yen ($200 million) to build. It would accommodate 170 jets per day (including SST's) and 5,400,000 passengers per year. The company would offer 150 percent value of the land currently held to anyone who would sell, plus generous cash bonuses to the scores of Sanrizuka farmers who were deeply in debt. Smooth, well-dressed Kodan representatives began paying personal visits to the peasant families. By fall of 1967, Kodan owned 80 percent of the site.

But on October 10, eleven days before the American march on the Pentagon, the Sanrizuka Hantai Domei decided to physically prevent a survey of the site. Calling for support from all of Japan, the farmers began sitting down at every conceivable access road. Kodan responded with two thousand riot police and a whole day of solid, unabashed beatings.

In the summer of 1968, the peasants waged two months of daily battles against survey teams, now under heavy police protection. They set barbed-wire barricades throughout the land and flung "golden bombs"—liquified human excrement, normally used as fertilizer—at the Kodan men with long-handled wooden dippers. By November, the Hantai Domei could rally eight thousand supporters.

By the summer of 1970, the resistance had built three full-scale fortresses on Kodan land, plus an underground tunnel system that could accommodate forty people with beds and a toilet.

In the fall, the police staged their first mass attack to protect a major survey. They were met with a barrage of golden bombs, rotten watermelons, burning tires, rocks and bamboo spears. Pitfalls trapped airport officials in business suits, and buckets of excrement fell out of trees onto passing riot squadrons. A number of farmers covered themselves with the stuff and attacked. Scattered Hantai Domei patrols ambushed airport officials who were trying to avoid the battle.

Finally the fortresses were destroyed. After three days of battle, more than sixty people were arrested. The survey was completed.

The following March, Kodan began the first of three forcible confiscations of land they could not buy. Pre-empting the local government, they brought in 3,500 riot police to confront a massive barricade guarded by 6,000 peasants and students flying banners such as *KEEP TO THE LAND UNTIL DEATH!*, *WE DEFEND LIFE, HEART AND FAMILY, WE WON'T GIVE AWAY THIS LAND, CULTIVATED WITH THE SOUL OF JAPANESE FARMERS,* and *THE DESTINY OF JAPAN DEPENDS ON THIS BATTLE!*

As the police moved in, a bulldozer was destroyed by a molotov cocktail. Throughout the months of preparations, heavy Kodan equipment had been disabled by sugar in gas tanks and other mechanical sabotage. Now it was one month before the airport was originally scheduled to open. Kodan finally took most—but not all—of the land it would need.

Through the summer, more pressure was applied to the last holdouts in a vain effort to avoid another violent scene. The resistance spent the time building massive barricades and digging an enlarged tunnel system.

On September 16, Kodan attacked with four thousand plainclothesmen and five thousand riot police clothed in bulletproof jackets and riding bulldozers, steam shovels, cranes, water cannons and armored trucks. There were five major resistance fortresses, three defended by peasants, two by students. They all had a varying number of towers growing out of them for observation and bombardment purposes, plus one sixty-foot broadcasting tower.

In defending their fortresses, the Hantai Domei destroyed eight bulldozers and at least one steam shovel. While the battle centered on the fortresses, roving bands of Hantai Domei guerrillas ambushed riot squads. According to the government, three groups beat three policemen to death at the Toho crossroads. In the afternoon, a huge mobile crane charged the Hantai Domei broadcasting tower and hooked onto it while Kodan police cut through with oxyacetylene torches. The tower finally crashed with eleven people still on it.

By the end of the day, the Kodan patrols were in control of all but two pieces of land, a graveyard and the small farm of a sixty-three year-old woman named Ohki Yone. She was among the very poorest of the peasants, and was essentially illiterate. In the midst of the fight she had someone write the following dictation on a board on her fence:

My land and home are going to be taken next, so I am going to fight like anything. When Kodan and the running dogs of the

*government come and trample me down with their bulldozers,
I'm still going to fight at this family grave with manure bags and
my late husband's sword.*

*From when I was seven, I was sent off to work as a nursemaid,
so ever since that time whatever I had to do alone, I did with all
my might. That's why the struggle has been the happiest time of
my life. It seems that my body—it belongs to me, but also it
doesn't belong to me. My body belongs to the Opposition
League. Over six years now I have been fighting with the League
and supporters. So no matter what anybody says, I'm going to
fight to the end.*

Kodan announced it would take Ohki Yone's land on September 21,
but came a day earlier for surprise. More than one thousand riot police
armed with three bulldozers and a steam shovel surrounded her land,
while Ohki Yone resolutely threshed her rice. As the heavy equipment
smashed through her fence and obliterated her house, Ohki-San
worked her grain undisturbed. Finally the police carried her off.
Nothing was left on her land to indicate she had ever lived there.

A week later, Sannomiyo Funio, a twenty-two-year-old member of
the Youth Action Brigade, committed suicide in protest against the
airport.

Kodan now had the land for the airport and soon it began construction. Over the next three years, Kodan constructed one four thousand-
meter runway and most of the facilities to accommodate passengers.

Meanwhile, the Hantai Domei built two structures of their own—a
thirty-meter tower and a sixty-meter tower. The first was built "for
practice," and the second was assembled and put up in two weeks in
1973, catching the government completely off guard. Standing within
a half-mile of the end of the runway, it effectively prevents any
airplanes from landing or taking off. As long as it is there, the airport
cannot open. It is guarded twenty-four hours a day by at least five
Youth Action Brigade members, who eat and sleep in an old city bus
parked at the tower's base.

The tower, two hundred-feet high, is firmly rooted in four massive
slabs of concrete. The iron structure is thick and solid, and the riveting
all looks professional. "The pieces were gathered in Tokyo," says
Koizumi, a leader of the Youth Brigade. "We brought them here
secretly and got the whole thing up in two weeks. Kodan never knew
what hit them."

The airport cannot operate with the tower there—it violates the civil air code, and being directly in front of where planes are scheduled to take off, poses a considerable hazard. About a third of the way up the structure is a small cabin—actually a plywood box with windows—in which the Hantai Domei die-hards plan to chain themselves when the crunch comes. "We will turn the tower into a human body," says Issaku Tomura, a leader of the Hantai Domei. "There will be people in the cabin and chained all up and down the tower."

Later that day, we find ourselves at Koizumi's house. He, his wife Miyo, and their two small children live in a three-room "struggle shack" on land that belonged to Grandma Ohki. Before she died, Grandma Ohki legally adopted the family so they could inherit her land and continue the struggle. She herself was buried at the site of the second runway "so her spirit could rise up and harass the airplanes."

Inside the house, as more members of the Youth Brigade arrive, the talk is of court cases.

The family and friends drink *sake* into the night. There's a dream-like air to this whole drama. Less than a mile away is a gigantic ultramodern machine that cannot operate because of a group of people who are supposed to count for nothing in the technological era.

☆☆☆

In the spring of 1978, as the American antinuclear movement approached take-off, we were once again drawn to Narita. A spectacular climax was drawing near, and the ecological fate of the archipelago seemed to hang in the balance:

Vietnam in Japan

Narita, Japan

SLOWLY, CAREFULLY, the manhole cover came off.

One by one, twenty saboteurs filed out of the sewer that had been their home since the previous day, leaving food, bedding and leaflets for the police to find later.

Above ground, the air was supercharged. A battle unparalleled in Japanese history was raging. Thirteen thousand riot police clad in heavy asbestos suits and carrying metal shields battled twenty thousand farmer and radical opponents of the New Tokyo Airport.

The date was March 26, 1978.

The police were hopelessly outfoxed. From a score of fortresses on farmland surrounding the airport, the Hantai Domei had staged bobs, weaves, and feints that had left the police bewildered. Organized into ideological and geographical cadres, signified by the color of their helmets and jackets, thousands of attackers maneuvered around fences and police lines, leaving the airport's defenses in disarray.

Suddenly, unexpectedly, the Hantai Domei moved into a frontal attack. Disciplined platoons of rock- and molotov cocktail-hurling radicals sent an armored vehicle crashing through a gate. Riot police, many of them young, inexperienced, and thoroughly frightened, scurried in all directions. It would be hours before they could regroup.

Meanwhile the twenty sewer dwellers dashed for a tall building at the very core of the airport. In the chaos of battle, the police had overlooked the nerve center of the entire operation—the billion-yen air traffic control tower, where complex and vulnerable computers sat with only a token guard.

From their perch on the sixteenth floor, a half-dozen busy technicians heard sounds of disruption below, but dismissed it as the product of workmen putting the final touches on lower-story support facilities.

A short-wave radio soon informed them of their mistake. They piled steel desks and chairs against the stairwell door and elevator, but it was already too late; the attackers were bashing their way through the tower's huge plate-glass windows, and entering from a small walkway that surrounded the tower.

The terrified technicians beat a frantic retreat through a hatch onto the roof, where police helicopters would rescue them two hours later.

The guerrillas who replaced them spent those two hours using lead pipes, sledge hammers, and axes to obliterate the television monitors, computer consoles, wiring systems, and huge plate-glass windows of the tower's top story. Downstairs, other cadre members ripped through charts and files, heaving them over a balcony and turning what had been a showcase of modern Japanese technology into a smouldering shambles. Outside, police watched helplessly as the anti-airport forces rampaged through the facilities.

Through the chaos and bloodshed, one thing had become quite clear—the New Tokyo International Airport would, once again, not open on schedule.

Nor would Prime Minister Takeo Fukuda, the fourth leader of Japan to risk his prestige on the Narita facility, be very happy. "It is regrettable," he told the world media. "I am sorry for foreign countries."

He also added that, because of this "serious challenge to law and order," the airport's tenth official opening would indeed be postponed so the tower could be rebuilt and so workers could weld shut the more than 350 sewer covers known to exist on the grounds.

Fukuda's unhappy view of the day's events was not universally shared. "There was great joy when the tower was smashed," says Muto Ichiyo, a noted Japanese writer. "People were laughing and laughing. Everyone was excited for weeks. Fukuda tied his fate to the airport and was made a fool of by millions. It was a great victory for the people of Japan."

Indeed, far more rode on those sledge hammers than the political career of Takeo Fukuda or the fate of a single airport. Narita had become both a symbol and top prize in the fight of Japanese farmers and ecologists to save what little remains of the archipelago's life support systems.

As the warfare at Narita escalated, the Hantai Domei welcomed the urban radical sects—but with strict conditions. The traditionally conservative farmers warned the students to leave their legendary factional fighting at home. No internecine warfare would be tolerated in the countryside. Any alliances between the farmers and the students would be built strictly on local control. Important strategy decisions would serve the needs of the area communities, dictated by local conditions. Nothing would happen without consensus from the neighborhoods around the airport.

On that basis, the coalition survived. Some urban student factions accepted the farmers' conditions, and have stayed with the fight for more than a decade, so long that most of them are no longer students at all. The media were stunned to discover that the "radical students" arrested in the March 26 tower raid included numerous civil servants and government employees.

That discovery heightened the shock of what had happened. The Fukuda government had put on its best face for the airport opening. Narita was not only the crowning achievement of Japanese technology, but also the nation's calling card to the world. A people that had opened its doors to foreigners only at gunpoint, now proudly lavished unlimited yen on an international doorstep. Narita was to be Japan's global face.

For the opening event, thousands of trinkets and commemorative medals and plates were prepared along with unrestrained media hype. Fukuda himself would attend the special Shinto ceremony, and the prestige of Japan Incorporated would be on the line as perhaps never before.

The spectacle of axes and sledge hammers smashing the brain of the airport could not have cut deeper. In a nation that barely has words to deal with "rebellion," the idea of such humiliation before the world, at the hands of its own, was nearly beyond comprehension.

Overnight the atmosphere in Japan changed. Suddenly, peasants and urban activists alike had a sense they could win. Chemical polluters, unwanted industrial expansion, atomic power plants—they all now seemed within reach of citizen action. "Our people are not used to fighting authority," says a professor active in the antinuclear campaign. "Sanrizuka has shown them it can be done."

For nearly two months the government made frantic preparations to try for an eleventh time to open the facility. Electronic equipment was repaired or replaced. Tens of thousands of helmeted riot police were stationed throughout the area. Arrests were made, Hantai Domei fighters passed in and out of jails, the area was constantly surveyed and searched.

On May 20, 1978, the airport officially opened. Ten thousand helmeted warriors snake-danced around the perimeter as the first commercial airliners set down on the single operating concrete strip. Thirteen thousand riot police guarded the interior.

Four days after the first planes touched down, we revisited the three-room farmhouse of Miyo and her husband, Koizumi, long-time mainstays of the Youth Brigade. Surrounded by their vegetable fields, the little house stood on land slated to become the still unbuilt Runway 2, in clear view of both the airport and two of the million-dollar motels built to serve it. The picture of the grass shack from which Grandma Ohki Yone had been dragged seven years earlier still graced the wall.

"I can't say we were 'happy' about what happened on March 26," says Koizumi, puffing on one of the "Peace" cigarettes you see sold in blue packs all over Japan. "It was a great victory, yes. But still our sisters and brothers are in jails and hospitals, and one of our people died.''

The jets outside continue to roar, but the night is warm and sweet. ''Do you ever get tired?'' I ask.

Koizumi sips some hot *sake*. ''No,'' he says. ''Never. We are really tough.''

They all laugh.

''No, really, we get really wiped away sometimes.''

''In the past history of Japan,'' says Koizumi, ''there is no case of successful nonviolence. The term for it here is touchy, confused with 'retreating.' Of course, I feel what I am doing is nonviolent because for me nonviolence is not a word of action but a way of thinking, the humanitarian approach. But in Japan it is not understood, because the liberation of human beings in a political sense is not understood in the heart.''

It is a difficult point to discuss, especially through a translator. We tell him that the Hantai Domei's propensity for throwing rocks and bottles at the police would result, in the States, in bullets from them, as at Kent State and Jackson State, where even that provocation was missing. They smile. ''Soon it might be like that here.''

Koizumi sits back. There is a silence. ''I can close my eyes and remember how beautiful it was here, before they built the airport. My main ambition,'' he adds, '' is to be a good farmer. The corporations' ambition is endless. But so is ours.''

As evening becomes night, a calico cat pads into the small room and settles under a low table around which we all sit. Misako and Hitori, the two kids, are stretched out in the next room, where Miyo and Koizumi also sleep. ''Our organization has no constitution,'' says Miyo. ''Nobody can really explain how it works, but it is organic and flexible. We have been together for a very long time.''

Dressed in a weatherworn blue sweater and traditional baggy blue peasant pants, Miyo talks through Mogi, our friend and translator from Osaka, where people are also fighting a new airport. ''We have lived seven years in Sanrizuka,'' says Miyo. ''Five years in this house. This is just the beginning of this struggle. I am very happy to be able to put energy into it.''

''We have no factions here,'' adds Koizumi with an air of finality. ''This is a real fight, not a game.''

In the morning, at his farmhouse-equipment store at the heart of Sanrizuka village, Issaku Tomura adds his wisdom of seventy years. Sporting a civilian beret (he wears a helmet at rallies), Tomura gestures at the howling aircraft. "So long as they fly in, the government will need thousands of troops."

Tomura moves with spry, wiry steps. Unlike many of the Hantai Domei fighters, Tomura's family has been in the area for generations. In the 1860s his grandfather was drafted into the Imperial army. While stationed at Yokohama, he met a priest who converted him to Christianity, a religion that has remained in the family. "I used to worship Christ in church, but that's not the way, because he was a man, an activist, a rebel. I used to praise him as the Son of God, but I praise him now because he was always conscious of the liberation of the suppressed. He was crucified not because Judas betrayed him, but because he defied state power."

Known for working weeks on end with four hours of sleep a night, Tomura seems to me like the eye of a hurricane. On May 20, we watched him deliver a speech to ten thousand helmeted, uniformed Hantai Domei fighters. His tones were even, his phrasing rhythmic, but he never shouted. After his talk, he sat calmly in a folding chair on the muddy grass, chatting with his fellow farmers.

Later, at his low, rambling house in the middle of Sanrizuka village, he talks of the reasons behind the struggle. "This is not just an airport they are building," he says. "This whole area, all the way from Tokyo, is to become a giant industrial city. The airport is just the first step. They want to destroy farming in Japan. Nuclear power plants, factories, dams—they take the water and pollute the land and make farming impossible. But how will we breathe? What will we eat?"

Tomura's yard is decorated with his scrap-iron sculptures, for which he is widely known. In the front of the house, at the street, is the family's farm-equipment store. Business, says Tomura, has not been good. "When we first started, our political consciousness was very low. We were just fighting for our community. But now we see the airport as just a small part of the larger struggle."

Nuclear power, he adds, will be the next big issue to rock Japan. "We do not have a real democracy here. We need to connect the dots between all the oppressed groups. In the space that is left, state power will be crushed."

Again and again Tomura and the media refer to this as "Japan's Vietnam." In a nation where deference to authority and conformity to communal codes reign supreme, Narita has been a volcano of rebellion. It has shown that citizens can rebel, can win, and, when defeated, can fight and win again.

As we talk, children and animals race through the garden. The phone rings constantly. Supporters dash in and out with the latest news from the front.

Still, tea is served, and the conversation is quiet. "The worst thing," says Tomura, "the hardest thing, was when they killed Higashiyama Kaoru." Higashiyama, twenty-seven, died after a police tear-gas cannister, fired at close range, struck him in the head. He was acting as a medic at the time. "The police do not behave like humans," Tomura says with a mixture of sadness and anger. "In their frustration at not being able to crush us, they go crazy."

I ask him the hardest question of all, how he feels now that the airport is open, whether it is not, as the media are saying, a defeat for the Hantai Domei.

He pauses a very long time. "We are learning to live together as a community," he says finally. "We need to establish a real commune, to share property and work together. The objective of our struggle is to make the image of a new world, new farmers, through the struggle.

"Many activists from many countries have lived in Sanrizuka village. This contact has helped us break up our parochial view of things.

"In an advanced nation, materialism is everything. For us, just getting back to nature is most important.

"As for the opening of the airport, it means nothing. The airport means nothing, whether they open it or not. The planes flying in mean nothing.

"Our struggle is like that of the Vietnamese. They fought for thirty years. And they won.

"We have fought together for thirteen years. We are proud of that history. We will fight together another seventeen years if necessary.

"We will win, too. We will fight as long as it takes. We will never give up."

PART THREE

On to the Sun

"If this were war, we'd have solar energy in a year."

—William Heronemous
Windmill Expert

9
Forging Alliances: Native Americans, Farmers, Workers

☆☆☆

Native Americans have played a seminal role in the American antinuclear campaign from its very start. New England tribal representatives were present at the first Seabrook rallies, and on Thanksgiving Day, 1976, Indian activists and nuclear opponents staged a joint rally at Plymouth Rock, one mile from the Pilgrim I atomic power plant. Throughout the West solar proponents have joined with Native American organizers to fight for reservation lands that have been pockmarked with uranium mines, coal fields, and electric generators designed to ship power to distant cities.

The Native American consciousness of the earth as an organic, spiritual being has also provided a basic and vital foundation for the prosolar movement. In the spring and early summer of 1978, feelings of solidarity and collective growth came together in the dramatic "Longest Walk," which spanned the continent, ending in Washington in a powerful demand for preservation of the few rights that remain to America's original inhabitants:

The Issue of Tribal Survival
(*New Age*, October, 1978)

They had forgotten the Earth was their mother. This could not be better than the old ways of my people. There was a prisoner's house on an island where the big water came up to the town. We saw that one day. Men pointed guns at the prisoners and made them move around like animals in a cage. This made me feel very sad, because my people too were penned up in islands, and maybe that was the way Wasichus were going to treat them.

In the spring it got warmer, but the Wasichus had even the grass penned up.

—Black Elk

I was born where there were no enclosures, and where everything drew a free breath.

—Ten Bears

ON JULY 15, 1978 THOUSANDS of Native Americans and their supporters marched through the streets of Washington, D.C. The action climaxed the "Longest Walk," a spectacular 3,300-mile trek from California to the East, an act of spiritual and political rejuvenation organized to help counter the continued official assault on Native American life.

Indians have been butchered and robbed since the arrival of the Spanish, and politicians have time and again declared them officially extinct. In recent years, the attack has been viciously renewed. More than one third of all native women of child-bearing age have been sterilized. Land, water, and treaty rights have been threatened with termination. Basic civil liberties on reservations have been drastically curtailed.

Though the enmity of white settlers toward native peoples dates back centuries, there may be a hidden force behind this latest armed charge—the headlong search for new sources of fossil and nuclear fuel.

In their haste to shunt native Americans onto the worst lands around, the nineteenth-century colonists unwittingly left the Indians sitting atop a huge share of this continent's geological fuel reserves. Roughly half the recoverable uranium in the United States lies under Dine (Navajo) land in the Grants uraniuim belt in northwestern New

Mexico. Many of the uranium miners in the area are native Americans, often working under primitive and dangerous conditions for as little as $1.60 per hour. Over the past three decades, more than 90 million tons of radioactive tailings have been piled up in this region, releasing horrifying quantities of deadly radon gas into the winds and inflicting cancer, leukemia, and birth defects on countless native people.

Critical uranium stores also lie under native lands in northern Saskatchewan. Huge uranium reserves face exploitation on aboriginal land in Australia. The largest uranium mine in the world sits on tribal land in Namibia, in southern Africa.

For western industrial powers, continued nuclear construction demands the removal or extinction of native peoples all over the globe.

Nor is atomic fuel all they're after. Gigantic coal reserves lie under reservation lands throughout the American West, involving as much as 30 percent of the country's recoverable coal. Nearly half the Crow and Northern Cheyenne reservations in southeastern Montana may soon be strip-mined out of existence. The Western Hemisphere's largest strip mine is being used to feed the notoriously dirty Four Corners power complex, in Navajo territory, and Peabody Coal is now strip-mining the Black Mesa in Arizona, sacred to the Navajo. Between corporate coal and uranium claims, a map of northwestern New Mexico Indian lands looks more like a war zone than an inhabitable region.

Nor is strip-mining the full extent of the devastation. The giant coal-fired Four Corners plant clouds the horizon in a region that once claimed the continent's cleanest air. The plant also devours billions of gallons of water a year, in a region where water is scarce. A half-dozen gasification plants planned for Navajo territory are slated to use more than 3 billion gallons of water per year each and a total of nearly 2 billion tons of coal over their twenty-five year lifespan. With $8–9 billion in construction costs at stake, the project is heightening local fears of increased corporate colonization.

Meanwhile, oil also has been found on numerous Indian lands, including the Pine Ridge, South Dakota reservation, where the Massacre of Wounded Knee (1890) was ultimately followed by a full-scale siege (1973).

That seige, in which two people died, renewed the native American's determination to fight back. Numerous tribes have since deter-

mined to block the devastation of their lands. Some, such as the
Wampanoag in Massachusetts, and the Passamaquoddy and Penob-
scot in Maine, are pursuing court actions that would return hundreds
of square miles to their original caretakers. Many tribes are now
working to establish a pan-national coalition aimed at regulating
native resources in a sane, systematic way.

The native revival underway is absolutely essential—for the survival of the continent as well as that of the tribes themselves, and their
way of life. No organized ethnic or national group inside U.S. borders
has suffered more unrelenting brutalization. Through the machina-
tions of the Department of Interior and the Bureau of Indian Affairs,
native American life has often been reduced to colonial servitude. The
FBI and its infiltration wing, Cointelpro, have consistently worked to
undermine the Indian political movement. In recent years, scores of
Indian activists have been framed, shot at, and murdered.

And now there are bills pending in Congress designed to finish the
job. House Resolution 9054, authored by Representative Jack Cunningham (R-Wash.), would terminate all treaty rights and reservations. Cynically labeled the "Indian Equal Opportunities Act," the
bill would deny all traditional water claims. (When Indians settled
with the military, they would always reserve their access to water,
recognizing it as essential to their continued existence.) The "Quanti-
fication of Federal Reserved Water Rights for Indian Reservations
Act," sponsored by Representative Lloyd Meeds (D-Wash.), states
that "all claims to aboriginal rights to the use of water are hereby
extinguished"—yet another repudiation of sworn treaties signed by
the U.S. government.

Crucial water disputes are already raging: they involve the commercial fishing industry in the Northwest, land developers in the
Southwest, and mining and power plant interests all over.

Meeds of Washington is also sponsoring the "Omnibus Indian
Jurisdiction Act," the purpose of which is to limit hunting and fishing
rights and to force the Indian nations to deal with individual states
rather than with the federal government. The bill would also curtail
the tribes' ability to regulate their own reservations.

And that's not all. The "Indian Trust Information Protection Act"
(SB 2773) would severely limit the information that individual Indians
could demand from the Bureau of Indian Affairs and other government offices.

In addition, a proposed section of the "Criminal Code Reform Act" (SB 1437) would invalidate all prior treaties between Indians and the U.S. This bill, sponsored by Ted Kennedy (D-Mass.), has drawn widespread criticism as one of the most comprehensive assaults on public civil liberties ever proposed—for Indians and non-Indians alike. The bill attacks, among other things, the right to demonstrate peaceably, to protest judicial proceedings, and to maintain silence before a grand jury.

The whole range of anti-Indian and anti-civil liberties legislation presents the frightening prospect of further unsavory compromises. If these bills fail, others will surely follow. "If we can't get the whole bucket," promises Representative Cunningham, "then we'll get it cup by cup."

The proposed legislation was the primary motivation behind the Longest Walk, but the action had strong spiritual overtones and important implications for the entire environmental movement.

The world has reached a "nuclear crossroads." Meanwhile, this continent's original spiritual ecologists are being driven toward the abyss.

Native tribes worshipped rather than exploited Mother Earth: there was never any question about their living in harmony with nature. Moreover, Indian writings and traditions have been extremely instrumental in prompting many non-natives to search for guidance from the planet and its natural spirits.

Building a native-environmentalist coalition, however, has been a complex process. Some groups counted among the ranks of environmentalism have in fact clashed seriously with native decisions. Thus far Indian and non-Indian activists have not always been sufficiently conscious of each other's origins and needs.

But the Longest Walk went a long way toward building the necessary understanding. And it came at a time when joint action has been rapidly developing. The four American reactor sites at which the most arrests have occurred—Seabrook, New Hampshire; Diablo Canyon, California; Trojan, Oregon; and Satsop, Washington—all sit on Indian burial grounds. Native speakers and organizers have been active in the antinuke movement from the very start. A joint antinuke/Native American action is now being planned for the spring in the uranium mining region of Grants, New Mexico.

Today, there is no separating environmental issues—nor nuclear

power—from Native American tribal survival: the ongoing anti-
Indian assault is as essential to the spread of atomic reactors and
weapons as anything that deals explicitly with nuclear fuel, wastes, or
siting.

It is also no exaggeration to say that harmonious human life cannot
exist on this continent unless justice is first guaranteed for the human
tribes whose traditional customs and culture uniquely embrace nature,
and whose claims upon the land, air, and water cut through to the very
soul of America.

☆☆☆

The Grants, New Mexico action took place in the spring of
1979. Involving some twelve hundred people, the demonstration brought many native tribes into a working alliance for the
first time in their long history, creating a unique pantribal
coalition aimed at fighting the destruction of their ancestral
homelands. The action also solidified working agreements between the Indians and nuclear opponents in the Southwest.

By then the push for the Meeds-Cunningham anti-Indian bills
that had helped prompt the Longest Walk had died down. But
parallel legislation will clearly continue to plague us all. As the
energy crunch worsens, we can expect escalating official assaults on the lives and land of native tribes.

By the spring of 1979 atomic power was also being widely
recognized as a major threat to our ability to feed ourselves. As
milk cows began to die in the vicinity of Three Mile Island, new
questions were being asked about radiation in our food chain.
Barrels of radioactive wastes, including plutonium, had long
since been discovered leaking at their Pacific Ocean dumping
ground off San Francisco Bay. Shortly after Three Mile Island,
alarming levels of plutonium were found in shellfish off the
Pilgrim I reactor at Plymouth.

Both fishermen and farmers have long been an integral part of
the antinuclear campaign. In the spring of 1979 farmers helped
kill a nuclear project at Tyrone, Wisconsin. The farm-Indian-
solar coalition that has grown up in central Minnesota has also

expanded to focus not only on local reactors, but on uranium mining in the Dakotas as well.

The underlying issue, however, is more complex than what atomic pollution can do directly to fishing and farming. Our entire food system has been put on a footing of spiraling energy use, trapping farmers and consumers alike in a bind of rising prices and falling productivity. In the winters of 1977–78 and 1978–79, the marches on Washington by the American Agricultural Movement seemed to underscore the dilemma—and the possibilities of an expanding farm-solar coalition. Key to the broadening of such a coalition, however, is an analysis of large-scale agribusiness and its effects on the patterns of energy use, on our ability to grow food, and on the people who actually live on and work the land:

Food and Energy
(*New Age*, April, 1979)

"Get big or get out."
—Earl Butz, Ralston-Purina Executive,
Former U.S. Secretary of Agriculture

OVER THE PAST FEW MONTHS, thousands of farmers have once again poured into Washington to demand changes in the way we deal with food.

Though the media has portrayed them as a cross between buffoons and bomb-throwers, the presence of American farmers marching, or driving tractors, in mass protest signifies a very deep and important ailment in how we feed ourselves.

Not surprisingly, the problem is also reflected in the thousands of people who have been marching against nuclear power.

For one of the biggest problems confronting American farmers today is the soaring cost of energy, energy upon which food production has become absolutely dependent throughout the industrial world.

At the same time, the emergence of modern agriculture as our number one consumer of petroleum (40 percent of the U.S. total), and one of our biggest users of electricity (2.5 percent of the U.S. total), has exacerbated our "energy crisis."

Farming with Oil

The two most important recent trends in American agriculture have been accelerated monopolization of land and increased mechanization.

In 1776 more than 90 percent of all white Americans lived on family-owned farms. By 1920 the urban-rural ratio was roughly fifty-fifty. Today less than one out of every twenty Americans lives on a farm. The ratio of two hundred years ago has been more than reversed.

Usually our history books present this story in the context of some natural and beneficial force of progress. The "miracle," they say, is that it now takes just one farmer to produce enough food for forty-eight people.

But the progression masks some harsh costs: land inequity, inadequate marketing systems, mass unemployment, ecological chaos, misallocation of energy resources, human suffering and profound doubt about where our food will come from in the future.

For modern industrial agriculture, as "productive" as it might seem, sits on the shaky foundations of high energy consumption and environmental exhaustion.

The basic force pushing people off the land and into the cities has been money. As control of American industry has become increasingly centralized, small farmers have been unable to survive in the marketplace. With credit, transport, equipment, seed, and other essentials thoroughly monopolized, farmers everywhere have fallen upon the mercy of bankers and merchants.

In its two centuries, the United States has never developed a food marketing or land tenure system that could guarantee a farmer's survival. A family could—and still can—work ten successful harvests only to be wiped out in the eleventh. Always in debt, never secure, American small farms have gone under at the rate of a hundred thousand per year since 1930, with farmers nearly always sent tramping off to the cities to become industrial laborers—and consumers of food shipped in from the land they once worked.

Their land then became the property of a handful of agribusiness giants such as General Foods, General Mills, Delmonte, Dole, Cargill, Ralston-Purina, and the like. Just 6 percent of the farms in the United States now control more than 50 percent of the farmland, and those corporations have inexorably substituted machinery—and cheap energy—for the departed farmers.

The giant tractor-combine that can plant, spray, and harvest ten thousand acres, owned by one outfit and growing a single crop, has become the cornerstone of our food supply, because agribusiness finds it cheaper and easier to buy equipment than to pay wages. The situation is exactly parallel to the use of electricity for automation in factories (see next article). Machines, with the aid of tax breaks and cheap energy, are cheaper than people and won't go on strike.

Chemical fertilizers, pesticides, and herbicides are also, essentially, labor replacements. The old ways of removing insects by hand, of weeding and cultivation with hoes and small tractors, and of fertilizing with manure are labor-intensive. Chemicals for pumping up the soil and poisoning pests and weeds are the agricultural equivalent of automation.

They also have a huge appetite for energy. By industry estimates, the production of chemical fertilizers alone requires eight hundred thousand barrels of petroleum per day. Pesticides and herbicides demand still more. Add to that the requirements of machinery, transportation, and processing, and you have a major cause of the "energy crisis." It now takes eighty gallons of gasoline to raise one acre of corn in this country. If American methods were applied world-wide, 80 percent of all energy would go to growing food.

Death of the Land

But the real productivity of the chemical pesticides, herbicides, and fertilizers is now questionable.

Insects still destroy one third of American crops. Some 250 strains of insects and mites are known to have developed pesticide immunity. Yet herbicide use has tripled since 1964, and according to researchers David Pimental and John Krummel, pesticides and herbicides now kill some 200 people per year, poisoning 14,000.

Meanwhile, chemical fertilizers have become an addictive drug for crops. The natural fertility of the soil comes from the organic matter that decomposes in it, and the earthworms and other organisms that live, fertilize, and die in it. Pumping in oil-based chemicals overwhelms the life in the soil and turns it into a dead medium for an ever-expanding petroleum "fix."

The impacted land then becomes prone to erosion, costing valuable topsoil and carrying the chemicals into lakes, streams, and seas, poisoning fish and people. By substituting a single nutrient, nitrogen, for natural balance, chemical fertilizers also leave us with food that is both nutritionally deficient and with a waste of energy. According to food researchers Carol and John Steinhart, more than 100,000 kilocalories of energy could be saved per acre merely by substituting animal manure for chemical fertilizers. Another 150,000 could be saved with crop rotation, and still more could be conserved in certain circumstances by using work animals instead of tractors. By improving on traditional methods, Chinese rice farmers have raised as much as 50 food energy units for a single unit of mechanical energy expended. In the United States the same 50 food energy units can cost from 250 to 500 units of mechanical energy.

In the 1960s, western technicians decided to export high-tech farming in the form of a "Green Revolution" based on "miracle" strains of crops and heavy inputs of chemical fertilizers, pesticides, herbicides, irrigation, and mechanization. Aiming for the highest output money could buy, the engineers of the "revolution" sent labor-saving devices into the areas of the world with the worst unemployment, thus further exacerbating both joblessness and urban crowding.

In addition, many of the new strains of seed were not naturally suited to the climates and land chosen, and basically drew their existence from energy-expensive irrigation and chemicals.

Because the Green Revolution was built on oil, the embargo of 1973 crippled it.

The chemicals that robbed the land of its natural fertility became too expensive to continue the "fix." Pesticides grew too costly and ineffective against insects that had developed immunity and now flourished free of their natural enemies. Irrigation and transportation costs soared.

And, most important, abandoning self-sufficiency left nation after nation more brutally undernourished than ever.

Single-Crop Insanity

Before agribusiness and the Green Revolution took hold, farming was generally done by shifting crops from one plot or field to another, year after year, with some land left periodically fallow. Because one year's crop replaced soil nutrients taken out the previous year, crop rotation strengthened the fertility of the land. By varying the location of their favorite ''meal,'' rotation also decreased the concentration of specific insects in one spot.

Most important, balanced cropping gave farmers and their region some safety margins. If one crop failed, a different variety planted in the next lot might well be thriving.

But the uniformity demanded by large-scale agribusiness has undercut that balance. Huge machinery and mass marketing have resulted in entire states, regions, and even nations being devoted to a single crop, regardless of the needs of the soil or the local people. Thus, for example, Ethiopia exported coffee and cocoa to Holland while thousands of its own people starved to death for lack of food, which could have been grown on the same cropland.

Undercutting local produce has also opened up the huge new industry of food packaging, which now ranks fourth among all industrial energy consumers, outstripped only by primary metals, chemicals, and petroleum refining. General Foods admits to producing more than *six billion* packages of individually wrapped foodstuffs in a single year. According to the Worldwatch Institute, a solid 36 percent of all energy expended in the food business goes for transportation and packaging, double what it takes to grow the food in the first place.

According to food scientist David Pimental and others, a pound of corn with 375 kilo-calories of food energy in it can consume some 450 kilo-calories of mechanical energy to grow—and then 760 kilo-calories for the tin can to sell it in, and then another 800 kilo-calories to get it home.

Commercial production of meat is an even worse energy waster.

Growing meat on marginal grasslands, with minimal labor inputs, can make a certain amount of sense in terms of food value returned on the energy and resources invested.

But agribusiness is another story altogether. According to Frances Moore Lappe's *Diet for a Small Planet*, commercial meat production now consumes up to 90 percent of all grains grown in the United States, fifteen pounds of vegetable protein per pound of commercial meat protein. Pimental adds that the production of 375 kilo-calories of meat on a feedlot can demand, under current methods, an incredible 29,000 fossil-fuel kilo-calories.

That production also translates into environmental damage. Because the cattle are fattened in overcrowded feedlots, they create huge piles of manure. The manure has enormous potential for both fertilizer and methane production. But commercial feedlots ignore this potential for methane, and the manure piles drain energy in the form of the labor and equipment it takes to move them out. They also become a pollution problem when rain washes their centralized impurities into rivers and the water table.

Food or Energy?

As a whole, agribusiness is both an energy waster and a destroyer of the land. Those statistics about how few people it takes to grow our food just don't take into account the thousands of people who build the machinery, the thousands who mine, treat, and transport the fuel, the thousands more who process, transport, and market the food once harvested.

In essence, agribusiness has moved people off the land only to put them to work at menial food-related tasks removed from the farm. Yesterday's farmer is today's cook at Burger Chef.

And yesterday's energy source is today's environmental nightmare. The centralized, high-technology energy supply promoted in part by modern agribusiness has in turn greatly contributed to a world-wide threat to cur ability to grow food.

For example, drilling for, transporting, and refining oil has of itself done severe damage to the food chain. Massive spills such as those of the *Torrey Canyon, Amoco Cadiz,* and *Argo Merchant* have threatened ocean fishing, as does routine leakage from offshore wells. Air pollution from refineries may affect long-range weather patterns, and with it our ability to farm. Giant hydroelectric reservoirs have inundated huge chunks of fertile farmland, as has unrestrained strip mining. Coal slurry projects threaten basic water supplies in key agricultural areas in the West, as have nuclear power projects all over the world.

The steam emissions from nuclear power plants raise still another fear, chemical pollution. Because the steam is thrown into the atmosphere through giant cooling towers, it must be chemically treated to kill micro-organisms which would otherwise clog the works. But the chemicals are not biodegradable. Passing through the towers, they enter the atmosphere and eventually settle on crops throughout the area.

At sea-cooled reactors, water sucked in from the ocean invariably carries with it some marine life, which is destroyed. The water then passes through the reactors to the sea, slightly radioactive and significantly hotter. It is also chemically treated, because the problem of keeping the pipes free of barnacles, jellyfish, and other potential clogging agents is the same as with the surface towers.

But aside from the chemicals and the radioactivity, heat is the real killer. Huge quantities of unnatural thermal energy can unbalance local marine ecologies, killing some species and driving others away. The industry has claimed that the heat will merely alter the local environment and attract new species. In fact, reactors are shut down as often as not. Instead of a steady stream of hot water, they produce a hot-cold-hot-cold effluent in which only the hardiest of species can survive.

One such creature is the shipworm, a bizarre sea parasite that has proliferated in Barnegat Bay, New Jersey, downriver from the Oyster Creek reactor. The shipworms are now eating the wooden piers and hulls in the bay, while crowding out other species, crippling the bay as a food source.

But all that pales next to the damage that could be done by a major radiation release. A graphic example is provided in California, where three reactor sites, San Onofre, Diablo Canyon, and Rancho Seco, offer the ultimate marriage of the dangers of centralized farming to those of centralized power generation.

The three plants, involving a total of six reactors, all sit in, or are upwind from, California's Central Valley, where much of America's winter fruit and vegetable supply is grown by agribusiness corporations. The lettuce, asparagus, broccoli, artichokes, peaches, grapes raised here constitute a significant proportion of the national supply.

An accident at Diablo, San Onofre, or Rancho Seco—aside from killing thousands of people—could eliminate a large percentage of America's winter food supply, and have an immediate and serious impact on the nation's ability to feed itself.

Small Is Plentiful

Nuclear opponents have now developed a "soft path" program for using solar energy, recycling, and conservation to supplant fossil and nuclear sources. A parallel solution is key to our crisis in food.

There are strong indications that chemical farming has reached a point of no return. U.S. Department of Agriculture statistics indicate that yields per acre in cereals have actually declined in America since 1972. Over-all growth rates in productivity have also declined. And fuel efficiency in producing at least one major corp—corn—dropped fully 24 percent from 1945 to 1970.

Switching to a cyclical system based on organic methods would both increase efficiency and decrease the fuel consumption that is exacerbating the world food crisis. Scientific composting (a methane producer), improved crop rotation, and the use of green manure (mulch and cover crops) all save energy while raising productivity.

It takes, for example, more than twice the energy to machine-spray with herbicides as it does to dig out weeds through cultivation, even by machine. Cultivating may not wipe out all the weeds, but it does aerate the soil. Leaving some weeds helps hold moisture, and adds important mineral balance and organic matter when the weeds decompose. (Some weeds, like lamb's quarters, are more nutritious than most crops anyway!)

Biodegradable pesticides, companion planting, and the introduction of natural pest-eaters such as ladybugs and praying mantids are excellent substitutes for energy-eating pesticides. According to Duane Chapman, writing in *Environment* magazine, the U.S. could expect to lose only 5 percent of its crops with an immediate abandonment of chemical farming. But the energy savings would be enormous. A study of twenty-eight corn-belt farms by the Center for the Biology of Natural Systems indicates that organic operations can produce comparable yields at comparable net profits while using half the energy or less.

But like solar power, organic farming involves more than a mere switch in technologies. Natural methods are labor-intensive. Because they rely more on human work than on capital and energy, they gravitate toward small units, where people live close to the land which they regularly work, and in which they have a personal stake. Organic farming makes the most sense where control of the land is decentralized, in the form of a patchwork of family- or community-run farms, where people are around to do work which large-scale machinery isn't really designed to handle.

For agribusiness, that's the ultimate threat. The whole thrust of corporate farming is to computerize food production, to concentrate ever-larger chunks of the planet into industrial units—chemicals, machines, energy. Small wonder the food giants use the same charge against organic farmers that the energy giants use against solar advocates: it isn't "practical."

And in their terms, they're right. Organic methods are not "practical" in terms of large-scale ownership and exploitation of the land.

The current methods, however, speak for themselves. Half the human race is now malnourished. Ten thousand people die each day from hunger. According to Pimental and Krummel, the U.S. lost 72 million acres of farmland to highways and urbanization alone from 1945 through 1972. Food prices in the U.S. are soaring at the rate of 10 percent and more per year. According to Uwe George, in *Deserts of the Earth*, the planet is losing 14 million farmable acres per year to deserts. The prognosis for both food and energy supplies under the current system is downright frightening.

The presence of protesting farmers in Washington carries a double message. Though most of them are not farming by organic methods, and are not energy self-sufficient, the small farmer in America does represent a last hope for some degree of decentralized land ownership. With that decentralization comes the hope for the spread of organic methods—and a solution to the energy crisis on the farm.

Our limitless, rural solar, wind, wood, biomass and methane resources, harvested by small farm units, could of themselves undercut any projected need for new industrial generating facilities, including nuclear power plants.

Statistics from the World Bank and other sources clearly indicate that small farm units produce more food per acre than big ones.

And organic methods ultimately have the best chance of flourishing in a nation where small farmers can survive. By buying organic produce, by starting to grow our own food, and by challenging the hold of agribusiness on the land, we can start to take some crucial strides toward solving both our food and our energy crises.

Farm organizations such as the National Farmers Union of Canada and local growers and farmworkers unions around the U.S. have already taken strong stands in opposition to atomic power development, both world-wide and in their own locales.

Non-farming opponents of atomic energy might now do well to work for the survival of the small farmer, and to recognize the use of organic methods on the land as an issue inseparable from that of nuclear proliferation.

Like solar energy, the land is infinitely capable of sustaining life, if people are given the chance to work it in a humane fashion, in concert with its organic needs.

And as in the struggle to stop nuclear reactors, the political barriers to toppling agribusiness are as big as they come, but not too big for a movement with the earth and the sun behind it.

☆☆☆

Coalitions with Native American, fishing, and farm groups in a sense flowed naturally from the rural and semirural base of the movement.

More difficult were questions of alliances with inner city and labor constituencies. The early antinuclear campaign was by and large lacking in participation from black, Chicano, and other minority groups, a fact amply pointed out by the major media. There were obvious reasons. "Seems like when we are sharing in the luxuries, the finer things in life, nobody questions where the welfare mothers are," said Dick Gregory at the 1978 Seabrook rally. "It's only when we get into adverse situations where they might bring in tanks, where they might be arrested, where everyone might be questioned. . . .I'll tell you one thing, you probably wouldn't be here neither if you had lost more loved ones by rat bites and lead poisoning and starvation than by atomic energy."

Three Mile Island, however, drove home the immediacy of the atomic issues to both the cities and the inner cities as no political demonstration ever had. By then other inroads had already been made, particularly with the introduction of solar technologies into urban neighborhoods in New York City, Detroit and southern California. Largely Mexican-American groups such as La Raza Unida of Texas, and the United Farm Workers of California, had also taken antinuclear stands. Minority coalitions on a larger scale would undoubtedly come with a stronger and more systematic introduction of solar technologies into urban and poor rural environments.

Coalitions with organized labor would also hinge on solar development in the urban scene. From the very start antinuclear and labor organizers were portrayed by the media as natural enemies. Things came to a head in June, 1977, when proponents of the Seabrook plant marched through the streets of Manchester, New Hampshire, shouting their support for the project (Chapter Five). Most of the marchers were utility and construction workers bused in by their companies. Their presence underscored the need of solar supporters to build some bridges to the labor movement, without which it seemed clear the leverage to stop atomic power would be difficult if not impossible to come by.

But there is clearly room for a meeting of the minds. The United Auto Workers (UAW) pioneered the first antinuclear intervention, taking the Fermi I fast-breeder reactor to the United States Supreme Court. Numerous local unions have registered opposition to nearby reactors.

The key to the issue remains the question of jobs. The nuclear industry portrays itself as a producer of both on-site construction jobs and of energy to keep the economy going.

Solar proponents argue otherwise:

Creating Jobs from Environmentalism
(*Mother Jones*, June, 1978)

"SOLAR POWER could be a much bigger jobs boom than the New Deal ever was," says Fred Branfman of the California Public Policy Center. Branfman, director of the CPPC, has just finished a major study—"Jobs from the Sun"—concluding that solar energy could create nearly 400,000 jobs a year in California alone between 1981 and 1990. It could also cut the state's unemployment rate nearly in half and save California about ten billion dollars in imported fuel, thus retaining capital and getting positive employment effects at the same time.

Environmentalists for Full Employment cites statistics indicating solar energy can produce up to *seven times* the jobs per dollar as nuclear energy.

Conservation, recycling, and pollution cleanup are also high on the list of job producers. According to forecasts by Chase Econometrics, a half-million jobs exist today in the U.S. as a net result of environmental legislation.

The jobs-per-dollar factor for solar energy, pollution cleanup, recycling, and the like is higher than for heavy industry because of the difference between "capital-intensive" and "labor-intensive" technologies.

Capital-intensive industries—such as oil refineries, reactors, heavy manufacturing—depend on expensive equipment, complex machinery, and monopolized fuels, all of which soak up huge quantities of energy and materials while relying little on human labor. A key factor in capital-intensivity has been electricity, a very specialized (and expensive) form of energy that has been used to power increasingly sophisticated equipment aimed at automation—i.e., the substitution of machines for people.

Labor-intensive technologies—such as renewable energy, recycling, and pollution control—rely less on material input than on human work. Because the jobs are simpler and decentralized, they are accessible to a large number of unskilled, semiskilled, and unemployed workers, both because work training is easy and because the jobs can be located in the neighborhoods and communities where people actually live.

But there are other questions beyond jobs-per-dollar output. Some unionists fear the decentralized industries may be increasingly difficult to organize. The relative simplicity of the work also raises fears that minimum wages will prevail, and that many workers may be forced to abandon hard-won skills and take pay cuts in the name of labor-intensivity.

To counter this line of thinking, Tom Hayden's Campaign for Economic Democracy, which has drawn up the SolarCal program for state-supported solar energy in California, wrote a proviso stipulating that priority for small-business loans be given to solar businesses hiring union labor. Jurisdictional agreements between the Sheetmetal Workers and the Plumbers have already been worked out in the solar-collector field. But what kinds of jobs a labor-intensive economy will produce, at what wages, and under what sort of organization are crucial questions still waiting for answers.

☆☆☆

Nuclear supporters in general also argue that the antireactor movement is proposing a "no-growth" economy, which to many unionists and others means unemployment and a rigid class system, whereby the rich stay rich and the rest stay where they are.

Nuclear opponents counter that solarization would offer a decentralized, broad-based economic expansion, built on stable energy supplies and prices, and a vastly expanded labor market.

They also point out that the atomic and utility industries have a long and dirty history of antilabor and antiunion activity. When it suits their political purposes, they build reactors with union labor. When it doesn't, as in four power plants being built in Texas, they use nonunion labor (which has also brought mainstream Texas unions into the antinuclear camp).

Overall, the prolabor campaign of the prosolar movement has had a powerful effect.

But a serious problem in many cases remains one of personal contact. Because of its strong rural roots, face-to-face conversation between antinuclear and union workers has been scant. But by the summer of 1978, labor committees within the various direct-action alliances, and independent groups such as Environmentalists for Full Employment, the Midwest Academy, the Democratic Socialist Organizing Committee, the Campaign for Economic Democracy, and numerous others could point to concrete successes in bridging the gap:

Unionizing Ecotopia
(*Mother Jones*, June, 1978)

IT WAS A DISCONCERTING VISION, a bitter reminder of the Vietnam era. Sure, there've been clashes between labor and environmentalists before. But none seemed quite so startling as the sight of three thousand hard-hats—many bused in and fed by public utilities—marching through the streets of Manchester, New Hampshire, chanting "Nukes! Nukes! Nukes!" Only weeks after the occupation by nuclear opponents of the site of the Seabrook reactors, the construction and utility workers stormed into Manchester demanding that in New Hampshire—and everywhere else—reactors be built, and built, and built.

This harsh demonstration followed hot on the heels of an angry twenty-three truck caravan of California loggers who toted an 8½-ton redwood tree carved in the shape of a peanut across the continent to Washington. "Enough of the damned Sierra Club," yelled the loggers. "It may be peanuts to you, but it's jobs to us." They wanted to halt expansion of the Redwood National Park in Humboldt and Del

Norte counties, where, as a result of park expansion, more than two thousand people would be out of work. Humboldt County, it was said, would be turned into an economic Death Valley.

In Maine, unions and environmentalists are tangling over the proposed huge Dickey-Lincoln Dam. Elsewhere they have fought over highways, mines, pipelines, refineries, offshore drilling rigs, you name it. With so much in their mutual interests, these two key American movements have spent too much of their time together at each other's throats.

But of late the talk has been of détente, and, in March, a major breakthrough occurred on the Redwood National Park front. AFL-CIO lobbyists, initially brought in by George Meany to try to stop park expansion, had begun to lobby for the park. What happened?

The unions finally came around on a $40-million rider to the Redwood bill—that's what happened. The rider, and other provisions, guarantees worker security and provides new park maintenance jobs. The compromise is a first, of sorts. And though construction unions still want to build nuclear power plants, the Redwood bill—imperfect though it may be—could stand as a landmark in the growth of a coalition with labor that environmentalists have long believed essential . . .and impossible.

The struggle between labor and environmentalists is a classic case of industry's pitting the have-nots against the have-nots. The movements would have never raised a placard against each other if workers were not forced to pay for their wages with pollution and environmental disease and, conversely, if environmentalists were not offered the image of dirty air and water as the price of jobs. The battle is built into the system.

Richard Grossman of Environmentalists for Full Employment, a Washington organization, founded in 1975 to join labor and environmentalists, puts it well: "If you don't have a full-employment economy, and no guaranteed job income, then working people are forced to carry the burden of environmental reform. The two issues are inseparable. And as that becomes clearer to people the two movements can't help but come together."

Recent environmentalist endorsements of the Labor Law Reform Bill—which would give unions broader organizing ability, especially in the nonunion South—and environmentalists' support of the J.P. Stevens boycott and the United Mine Workers' coal strike haven't hurt, either.

But the thaw has just begun. The gaps in understanding between these two movements are still there, as a visit to union headquarters amply shows.

First, There's Food

Rudy Oswald, director of research at the AFL-CIO, lays out the problem. "Everyone is concerned with the general state of the environment. But our people are still dependent on jobs for income, and we have to wonder if some people aren't against every project, everywhere."

A smiling, well-trimmed man in his 40s, Oswald keeps his offices in the labyrinthine AFL-CIO national headquarters on 16th Street in downtown Washington. Like many of his counterparts in the union bureaucracy, Oswald is dubious about ecological ideology. "I'm not sure where environmentalists are at. Are they really for getting rid of the steel industry? The difference for us is between getting rid of it, and producing steel in a more healthful way."

In fact, the union movement cannot be fairly labeled either monolithic or anti-ecology. Labor support was essential, for example, in the passage of Clean Air, Clean Water, Toxic Substances Control and other key environmental legislation. Some of the thornier environmental questions have split the diverse group of constituencies that make up the unions:

• The national AFL-CIO opposes a nationwide bottle bill, but many locals support such laws in their home states. The national supports nuclear power across the board, but several of the major industrial unions that make up the AFL support cutbacks in the breeder-reactor program, and some locals are fighting power plants in their own areas.
• The United Auto Workers (UAW)—not part of the AFL—has backed the car industry in opposing auto-emission standards, but it also helped bring the nation's first major antinuclear legal intervention all the way to the Supreme Court.

Today UAW locals are split on atomic energy, but the international has been solid on other environmental issues.
• The Sheetmetal Workers' International also sees a bright future in assembling and installing solar collectors; it has eight organizers and a film (*Under the Sun*) in circulation pushing solar power. William Winpisinger, the ebullient president of the Machinists International,

sits on the board of Sun Day with the Sheetmetal Workers' Ed Carlough and the UAW's Doug Fraser.

• Coal miners have opposed nuclear power reactors, as well as a dam project that threatened to flood a recreation area used by miners.

• Railway and transport workers are fighting Mississippi lock and dam projects that not only would uproot local residents and unbalance the environment, but also would pave the way for more federally "subsidized" barge transport, which would undercut their jobs.

Rudy Oswald comes back to the point: "There are a wide variety of issues we can agree on, no doubt about it," he says. "But we can't settle for just the environment. The environment won't do you any good if you haven't got anything to eat for dinner."

The No-Growth Nightmare

In the worst of times, saving the environment has brought direct hardship to labor: "We have had cases where dams were being built and where our members have gone a thousand miles to go on the job," J. C. Turner, president of the Operating Engineers, told a recent Washington conference. "They've put their kids in school in September, and then in December there has been an injunction. The injunction means the shutting down of the dam or whatever the job is, and, of course, they have usually used up whatever money they had to get there and get going. Then they find themselves in the winter and the middle of the school term with no job and no income."

Such events—far too numerous—have gone a long way toward taking hard cash out of union treasuries, bread and butter off rank-and-file tables and good will out of working people's hearts. And fault of the system or not, these sorts of dislocations have led a good many unionists to draw some conclusions of their own.

"It's arrogant for hard-core environmentalists to say they care more for the world they're leaving to their kids than we do," charges Reese Hammond, education director of the Operating Engineers. "But if it's 'no growth' they're advocating, then what they're really saying is: 'We've got enough for ourselves, but you stay down there.' "

"No growth" is a red flag that pops up frequently in union halls and offices. It epitomizes a fear that environmental concerns might be a cloak for a sort of militant "Waldenism" aimed at returning America to some romantic pre-industrial era—at the expense of the jobs and expectations of working people.

"We've been boxed into that corner too often," complains Anna Gyorgy, an organizer for the anti-nuclear Clamshell Alliance. Clamshell, along with other environmental groups recently came out in support of the Labor Law Reform Bill and has worked closely with mine workers. "We're not no-growth," she adds, "what we advocate is a different kind of growth, one that's equitable and environmentally sound."

Affable and cigar chomping, Reese Hammond is familiar with these arguments, but he still differs on important specifics. Hammond's International Operating Engineers (IOE) is a 415,000-member organization that counts 300,000 construction workers on its rolls. On the table between us are four slick copies of the *International Operating Engineer: A Magazine of Technical Progress*. The cover story on the top issue is an anguished, strident tale of the Clamshell Alliance and the history of the Seabrook reactor, where 20 percent of the final construction work is scheduled to be handled by the IOE. The story is called "Sorting Out the Seabrook Snafu," a reminder that the road to coalition on the nuclear issue will be a rocky one indeed.

View from the Kitchen

Laborers' International's Jim Sheets is a huge, intense, gregarious man. His is a giant construction union, representing more than half a million construction workers in the U.S. and Canada. Jim Sheets, the research director, and construction workers in his union may have the most to lose from environmentalism in the short run.

"Our welfare is tied very intimately to the developing economy," he told me in the library of the International's Washington office. "Which is why we have to part company with the environmentalists fairly frequently." Construction workers remain staunchly pronuclear, as Sheets would readily admit. But sometimes labor's frustration with the environmentalists may stem not from any one particular issue, but from trying to deal with a movement that is neither unified nor fully tangible.

"It comes time to build something," Sheets says with a trace of irony in his voice. "So we say, 'Okay, let's build coal.' And then another damn group comes out of the thicket and says 'Go build it in somebody else's back yard, and send the power to us.' As far as I'm concerned, a lot of these so-called environmentalists are a bunch of elitists who don't want the view from their kitchen window messed up."

Gail Daneker at Environmentalists for Full Employment has heard these sentiments often. "Labor works by giving a little here, getting a little there," she says. "But there's no way you can get just a little nuclear power. You also find some compromises that the Sierra Club will go along with, only to have Friends of the Earth turn around and sue the hell out of you."

Of such stuff bad feelings are made. Furthermore, there seems little incentive for unions to coalesce with ad hoc groups that may have little to offer in return for the compromise, and may not even be around once the deal is struck. "The labor movement is there, and it's gonna be there from here on out," says Jack Sheehan, a lobbyist for the 1.2 million-member Steelworkers' Union, and a key figure in the passage of national Clean Air and Clean Water legislation. "But the environmental movement gets a strong leader or a specific issue, and then people move away, the whole thing folds up, and whatever support you were hoping to get is gone." Despite these clashes—real and imagined—of origin, structure and lifestyle, a backlog of common ground is forming. "The labor movement owes environmentalists a great debt," says Frank Wallick of the UAW. "Without Earth Day and the research that's come out of environmentalism, we would be far behind where we are today in occupational health.

"As for the rest of it," he continues, "you must remember that we have to worry about what's gonna happen to people in *this* plant, with *this* contract and *these* fringe benefits. For all its faults, the union movement is still based on people who are trying to get a paycheck every week."

Evolutionary Labor

Indeed, with some 22 million dues-paying members, the American labor movement is bigger than many countries. Its core, the AFL—CIO, counts some 13.6 million members organized into 106 national unions and 62,000 locals conducting about half a million meetings a year. It also has a $110,000-per-year president who boasts of never having walked a picket line or led a strike.

As a social movement, American organized labor hosts two warring factions. One is the conservative and often corrupt "tuxedo unionism," in which the leadership identifies more strongly with management—politically and in lifestyle—than with the rank and file. The other is the working-class militancy that gave the movement its birth.

The division dates back at least a century, when the unions were born in brutal struggles that killed hundreds each year. Personified by the legendary Eugene V. Debs, the rank and file tended toward industry- or class-wide organization that would fight not only for wages and working conditions, but also for a new society. Their vanguard became the revolutionary Industrial Workers of the World (Wobblies), a group dedicated to the overthrow of capitalism.

But as industrialism advanced, a skilled and better-paid elite gravitated toward trade associations, organized by specific craft and concerned almost exclusively with wages and working conditions. Their leader, Samuel Gompers, viewed the fledgling American Federation of Labor's craft unions as "business organizations." As such, many of them regularly excluded blacks, women, and the unskilled, and some even helped break the strikes of rival unions.

As World War I approached, the internal conflicts sharpened, and Gompers struck a deal with Woodrow Wilson. Though the majority of American workers bitterly opposed U.S. intervention in the international conflict, Gompers joined with Wilson's war Cabinet in exchange for official recognition of the AFL—and for an all-out attack on the labor leftists.

War thus brought prosperity, prestige, and power to the AFL, while Wilson used it as a cover to smash the working class Left. Gompers' rivals inside the AFL were purged, Debs was imprisoned and the relatively powerful Socialist Party was bombed, burned, and jailed out of existence. Wobbly organizers were harassed, deported, and murdered. Massive postwar strikes were crushed, and a well-choreographed Red Scare sent thousands of labor radicals underground or into prison, leaving Gompers virtually alone atop a sanitized movement.

But the 100-percent-Americanism that followed in the twenties simply buried the conservative and increasingly timid union mainstream. A national swing to the right gave industry a clear upper hand, and, without the moral fervor and organizing energy of the purged leftists, unionism declined.

Depression brought back the whole cycle. Grass-roots radicals and the fledgling Congress of Industrial Organizations (CIO) took up where the Socialists and Wobblies had left off in organizing the unskilled, women, minorities, and the unemployed. Sit-downs—an old Wobbly tactic—and other radical tools flew back into the fray against Ford, Big Steel and King Coal.

But World War II brought more alliances—and more purges. Despite the defiant strikes of John L. Lewis's mine workers, labor fell into line for war, and in its wake the CIO moved in with the AFL.

By the sixties, the core of the AFL-CIO was in the harsh but aging hands of George Meany, a Gomperesque plumber from Brooklyn who may be remembered chiefly for his support of the Vietnam War. That support was echoed in the streets by hard-hats from the construction trades, who make up the conservative heart of the AFL mainstream.

But despite the conservatism at its core, American organized labor still faces strong and often violent opposition from industry. The Sunbelt remains largely open shop, and the Right is now spearheading a massive, well-financed national campaign against unionization. Last year the AFL lost nearly half a million members, and many more are now finding it prudent to "put their union cards in their shoes."

Meanwhile, labor's clout on Capitol Hill and in the state legislatures has slipped. Much of the blame has recently been laid on union leadership, particularly within the AFL. Almost totally male-dominated, the old-line Meany regime has drawn growing fire for being out of touch with the rank and file, for its apparent unwillingness to organize new industries, the unskilled or the unemployed, and for its Cold War politics and close identification with industrial management. The AFL has yet to regain many natural progressive allies lost during Vietnam. It has been difficult, for example, for traditional leftists to sympathize with the unions' cries about unfair competition with the cheap labor of Hong Kong, Taiwan, and Korea when the AFL supported a war aimed at turning Vietnam into the same kind of cheap-labor camp.

Inside the unions, however, a new militancy seems to be on the rise. Peaking during the great coal miners' strike this year, the rank-and-file ground swell has brought a rash of wildcat strikes, strong reform movements within the Teamsters and Steelworkers and a rising opposition to Meany within the AFL, a sentiment strongly identified with the Machinists' prosolar William Winpisinger.

The direction organized labor will take "after Meany"—or even while he's still around—is anybody's guess. But few doubt that labor is under strong attack, on the brink of a major upheaval, and that it could use some allies.

Environmental Employment

Richard Grossman's voice has a strong hint of a Long Island accent. "It's a classic case of Big Money's saying, 'Go ahead, you two fight it out.' But if you look at the record, you'll see that these same right-wingers who are the first to scream about jobs also have terrible records on both union and environmental issues. Big Money really couldn't care less about working people or the environment."

Tall and muscular, Grossman works with Gail Daneker out of the poster-laden offices of Environmentalists for Full Employment, perched on a third floor in downtown Washington.

Last year, Grossman and Daneker co-authored and published a pamphlet called "Jobs and Energy," 25,000 copies of which have been distributed to people in the two movements. Among other things, the pamphlet contends that solar energy could create many times more jobs than nuclear or fossil energy and that the main financial attraction of electricity has been its ability to power machines that put people *out* of work. Currently EFFE is working on a second study, "Jobs and the Environment."

"Our job has been as much to educate the environmental groups as it's been to talk to labor," says Gail Daneker. "People have got to be willing to take on the economic and employment issues. Whether it's the redwoods or nuclear power or cans and bottles, it's also people's livelihoods we're dealing with."

Thin, dark-eyed, and energetic, Daneker was instrumental in pulling together forty-six prominent environmentalists to endorse the Labor Law Reform Bill, a "must-win" bill for the AFL-CIO. The bill may come to the Senate floor this month, where it faces strong opposition from industry—particularly that located in the nonunion Sunbelt. Utah Republican Orrin Hatch has promised a filibuster on the bill, which he calls "a reprehensible monster." The act would expand the National Labor Relations Board, grant back-pay awards of up to 150 percent to workers judged to have been fired for union activity and allow union organizers equal time to address employees required by management to attend antiunion meetings. Since Daneker organized environmentalist support for the crucial bill, the AFL has taken grateful note.

But that doesn't mean Daneker and the unions see eye to eye on everything. "You talk about what you can agree on," she says, pushing aside her long black hair. "Right now, we can't talk to most of the unions about nuclear power, not even the prosolar ones. But jobs on the local level—yes. Jobs from solar energy—yes. Jobs from environmental cleanup—yes."

A basic theme Grossman and Daneker have been developing is that environmental cleanup creates employment, an argument given substantial weight by the Environmental Protection Agency (EPA).

Since 1971, the EPA has kept a running count of jobs lost through ecology crackdowns. Their figures show that pollution-control standards have contributed to the loss of about 21,000 jobs, with another 24,000 currently in jeopardy. The agency concedes that the figures may be on the short side, since they don't account for plants with fewer than twenty-five workers.

But in virtually every case on record, the factories that had to shut down were marginal facilities with other financial difficulties. A classic example of a plant that was going down anyway was the Lykes-Youngstown Steel Mill, where the parent company screamed bloody murder about environmental controls, but had milked the factory assets dry and let the machinery deteriorate. "A firm doesn't close just because it has to put in a scrubber," says EPA's Anne Smith. "The regulations may contribute to the downfall of a doomed industry. But rarely, if ever, are they the sole cause."

The EPA now points to a booming pollution-control industry that comes with a $15-billion federal, state, and local budget, and that has put at least 300,000 people to work in pollution control, and perhaps as many as two million in the entire economy. "All the studies indicate that environmental concerns have a positive impact on the economy and produce a net increase in employment," says Smith. "It's our feeling that pollution control favors labor."

Indeed, while factory owners bleat that cleaning up puts people out of work, it's usually their own dislike of paying wages that causes the problem in the first place. "It is possible, and industry knows it's possible, to build a new turn-key plant that will not pollute," charges Tony Mazzocchi of the Oil, Chemical and Atomic Workers International Union (OCAW). "But the real problem starts when industry won't hire the necessary personnel to run the place right. The machinery deteriorates—and pollutes—and then the community and the workers have to pay for it."

Mazzocchi should know. As a key organizer for the OCAW, the aggressive Mazzocchi helped lead a landmark 1973 national strike against Shell Oil on issues not of wages, but of worker safety and environmental control. Eleven national environmental groups supported the five thousand Shell workers who walked the picket lines for five months. Now an OCAW vice president, Mazzocchi has been a key figure in the move to bring labor and environmentalists closer.

"Corporate America has painted everyone into a classic corner," says Mazzocchi. "Now it's jobs versus the environment. The worker has a choice between earning his livelihood and dying of cancer."

Ballgame of the Eighties

There is more to environmentalism than pollution at the factory, though. Central to it is the idea that massive-scale technology has reached its point of no return, that the planet is too small and the natural balance too fragile to allow industrial expansion to continue out of control. Cleaning up the workplace treats a serious symptom, but the core of the conflict comes from the capital-intensivity of industry—overcentralization of resources and authority, and a spiraling dependency on expensive machinery, exotic materials and unlimited energy consumption. All of which the environment, and labor, can no longer afford.

As the work of Amory Lovins, Barry Commoner, the EFFE people, and a host of others has shown, an economy built on recycling, renewable energy, and a decentralization of resources would preserve the natural environment and at the same time create more jobs.

But hidden away in the jargon are some time-honored questions. How will income be distributed? Who will control industry? "Without democratic control of the workplace, you could have both solar energy and fascism," says Tony Mazzocchi. "My main concern is that people make their own decisions."

An attack on "capital-intensivity" is clearly an attack on capital—and the planet's power structure. Decentralization of energy could open the door for democratization of the workplace and community control of resources, never big favorites at multinational headquarters. Simplifying technologies and using less energy flies right in the face of a well-established technological elite . . .and the world's biggest monopolies. And the establishment of job security means undercutting the labor surplus that has guaranteed factory owners a cheap, mobile work force ever since industry took off in this country.

These are, to say the least, basic issues. "It's not clear if people are really ready to take on the employment question," says Richard Grossman. "But when you fight for full employment, or for an end to militarism, or an end to nuclear power, you've ultimately got to face the question of who really owns the resources in this society."

For all the faults of the American labor movement, unionization has been the vital link to maintaining any semblance of worker rights in this society. And unionization has caught this new social push, this campaign to save the natural environment, on the horns of a dilemma. Short of winning everyone's right to a job and a decent income, those who would make major changes in the industrial system must also be able to guarantee the needs of people displaced by the change. At the very least, they will require the support of organized labor. And labor has been slow to come around precisely because of well-founded fears that its members will suffer most in this shuffle to reorder the economy.

The major breakthroughs in the reordering have come precisely where both criteria have been met—where the jobs and income of redwood loggers were guaranteed in exchange for expansion of the park, where environmentalists could do more than merely illustrate how ecological concerns meshed with the needs of people to make a living—but actually had the clout to make it happen.

And therein lies the key to what may be the most important political ballgame of the eighties. "I've never met a worker who wanted to work in a polluted plant or live in a polluted environment," says Tony Mazzocchi with some bitterness. "People have to start understanding that workers bear the costs of environmental illness as well as the economic costs of shutdowns. If we intend to stop a project from being built, or from continuing to operate, the first demand that has to be laid down is that all workers are paid full pay for life. And I can tell you, you'd better stand out of the way as those workers trample out of that plant, if they are guaranteed full pay for life."

☆☆☆

The Green Bans
(Mother Jones, June, 1978)

A UNIQUE AND IMPORTANT COALITION between environmentalists and labor unions has come about in Australia, in a movement known as "the Green Bans."

Spurred on by Jack Mundey, the squat, powerful former leader of Australia's Builders' Labourers union, the Green Bans have resulted in cancellation of more than three billion dollars' worth of environmentally destructive building projects. Essentially the Green Ban is a labor veto of socially and ecologically unsound plans. Mundey devised the strategy when a developer tried to move in on Kelly's Bush, the last remaining piece of Australia's bushland in Sydney. When local residents tried to stop the development and failed, they turned to Mundey's union. After careful consideration and a public meeting, the union posted a Green Ban, refusing to build on Kelly's Bush.

The developer then threatened to use nonunion labor. Mundey replied that the Builders' Labourers would then leave some half-finished office buildings owned by the same developer as a monument to the virgin land. The developer gave in; Kelly's Bush remains intact.

Since then more than a dozen other unions have participated in Green Ban designations, and Mundey claims that at least 25,000 Australian homes have been saved from death by freeway.

Australian unions also regularly blocked troop and weapons shipments to Vietnam, and today they are engaged in a fight against the mining and shipping of uranium. Australia has no commercial nuclear-power plants, but the unions are wary of radioactivity from mining and mill tailings.

With a long tradition of radical union organizing, Australia is 60 percent unionized. The workers enjoy strong community support for their Green Ban decisions, of which there have been more than forty. "The unions have changed the whole thinking of environmentalists into a more urban reality," says Jack Mundey. "We want to build buildings that people need."

10
The Promise and Threat
of Solar Power

☆☆☆

The whole debate over nuclear energy has long been posited on two basic assumptions—that nuclear power was cheap, and that it was irreplaceable.

Neither is true.

For starters, the capital costs of building atomic reactors are out of control. The price for building a new reactor has quintupled since 1970 and is expected to triple again by 1990. By the summer of 1979 the official cost estimate of Seabrook Nuclear Power Station had soared to $2.5 billion. Independent experts put it closer to $4 billion, more than four times the original estimate, and an astonishing $20,000 per New Hampshire household.

Throughout the debate, the nuclear industry has complained that front-end cost increases have been caused by citizen interventions. The legal challenges and delays, they say, have unfairly forced up nuclear costs.

But those interventions, and the safety and environmental improvements they forced, may have already saved us from numerous Three Mile Islands, and may save us from more in the future.

As for the industry, the bulk of the increases have come not from legal delays but from the soaring costs of materials, labor, and financing—and from the producers' own miscalculations. The early low cost estimates were pure guesswork, based on little other than wishful thinking. As the Atomic Industrial Forum, a pronuclear consortium, has put it, "estimating capital costs for power plants is like shooting at a moving target."

By the summer of 1979, atomic energy had been all but priced out of the market, on the basis of initial construction costs alone. According to a study prepared by a New York energy economist, Charles Komonoff, the cost of an installed nuclear kilowatt will be $2000 by the year 1986; the cost of building coal will be half that.

Komonoff's study was done before Three Mile Island. That accident, he says, "has pushed nuclear power beyond the brink of economic acceptability. The nuclear versus coal debate is over but for the shouting."

Fuel Costs

And if construction costs have gone beyond the pale, so has the price of nuclear fuel. Originally, the chief attraction of atomic energy was that its fuel costs were to be a mere fraction of the over-all operating costs of a reactor. Uranium was thought to exist in abundance, and theoretically reactors would use so little of it as to make fuel a negligible cost input.

But in recent years uranium had soared from $7 per pound to $40 and more. Some experts have warned that it could go high enough to make the price of fueling an atomic reactor comparable to that of fueling a coal-fired plant (which costs far less to build). Part of the reason is that uranium is even more thoroughly monopolized than fossil fuels, its price controlled by an alliance of cartels now widely known as "UPEC." The cartel is so strong, in fact, that it recently forced the mammoth Westinghouse corporation to renege on fuel supply contracts it had made with a score of American utilities, resulting in a long string of multimillion-dollar lawsuits. The case led Westinghouse to bitterly accuse Gulf Oil and other energy cartels of monopolizing uranium fuel, but even giant Westinghouse was in the end unable to defeat them.

The fuel supply problem has been exacerbated by the fact that reactors are burning up the uranium faster than was originally planned. The crisis has been deepened by the enormous dollar and energy cost of facilities designed to enrich uranium fuel. The Portsmouth enrichment facility on the Ohio River consumes as much electricity as does the city of Cleveland. Overall, uranium enrichment consumes some 3 percent of the national electrical supply, fully a quarter of the electricity produced by the national nuclear program. Meanwhile, *Business Week* has predicted enrichment costs could rise by $30 billion in the next fifteen years.

None of this will add an ounce to the actual amounts of uranium available. Uranium exists in a finite amount, just like fossil fuels. There may be barely enough uranium to operate the power plants we now have for their projected life cycles. That means prices, in the not-too-distant future, will skyrocket again, much as fossil fuel prices have.

To avoid that, the nuclear industry has long promoted the so-called breeder reactor, designed to produce more fuel than it uses. One prototype was the Fermi reactor that almost blew up south of Detroit in 1966. Another is the Clinch River Breeder Reactor Plant in Tennessee. The Carter Administration has opposed this plant on environmental and technical grounds. Critics charge it is a poorly designed, poorly constructed "turkey." The industry, however, is desperately holding onto it—and the breeder program as a whole—in hopes it can deliver them from the inevitable radioactive fuel shortages.

But breeder technology remains very much an experimental fantasy. At best it's a dangerous, marginal, fantastically expensive experiment with no guarantees; at worst it could bring us yet another variety of nuclear disaster.

Poor Performance

Meanwhile, the commercial plants have been plagued by accidents, shutdowns, miscalculations, equipment failure, operator error, and by constant alterations to meet new health, safety, ecological, and legal requirements. Years ago the industry maintained that the average nuclear power reactor would generate at 80 percent of capacity. But the U.S. program as a whole has operated at roughly 60 percent of capacity. Says the *Wall Street Journal* of American reactors: "Their unreliability is becoming one of their most dependable factors."

Yet behind that "dependable unreliability" stands billions of dollars in federal—taxpayer—subsidies. Ever since the Manhattan Project, research and development for the nuclear industry has been billed almost entirely to the taxpayer. Insurance is still a federal responsibility, as is enrichment, safety monitoring, and plant security.

And none of that can hold a candle to the as-yet inestimable costs of waste disposal and decommissioning of plants.

All nuclear power plants generate huge quantities of radioactive trash, and we really don't know what to do with any of it. One early plan was to recycle—"reprocess"—spent fuel rods, turning some of their radioactive components back into usable fuel. Three plants have been built to do that. The first opened at West Valley, New York, in 1966. It closed under AEC order in 1972, never to reopen. It now houses some 600,000 gallons of high-level radioactive wastes whose final disposal will cost a minimum of $600 million, and possibly much more.

A second reprocessing facility, built by General Electric at Morris, Illinois, never operated and has been abandoned except as a temporary waste storage dump. A third—the $250 million plant at Barnwell, South Carolina—failed to meet federal standards. Its builders are now petitioning the government for a $750 million bailout.

Meanwhile atomic wastes are piling up at reactor sites throughout the country. Nobody knows what to do with them, and no one can even begin to estimate how much it will cost to solve the problem. There are few who doubt the final price tag will be astronomical, or that it will be the taxpayers who will be stuck with the bill.

As for the reactors themselves, if they actually operate for the thirty to forty years they were meant to operate, they'll have become far too radioactive just to let sit. Some pieces will have to be carted away, others buried, still others fenced and guarded virtually forever. Most utilities now estimate the cost of doing all that at $100 million per reactor. But the price could actually run into the billions.

Economic Waterloo

Beyond even all that is the ultimate "hidden" economic risk of nuclear power, a risk that became a little less hidden at Three Mile Island.

It's always been known that a nuclear reactor accident could be incredibly costly. The one at Fermi in 1966 eventually led to the scrapping of that $133 million facility—a loss that still figures in the bills of Detroit Edison ratepayers. The fire at Brown's Ferry cost customers of the Tennessee Valley Authority $150 million.

The costs of cleaning up Three Mile Island could run into the hundreds of millions. The reactor itself may ultimately have to be scrapped, after only three months of sporadic operation, costing the ratepayers of Pennsylvania and New Jersey a billion dollars. The destructive economic impact of the accident on local farms and business will add still more.

But consider this: at the time of TMI, the United States depended on nuclear power for 10 to 12 percent of its electricity. Atomic power thus remained a marginal commodity—though some areas, such as Chicago, New England, and the Carolinas, were more dependent than others.

But give the United States, as some hope, 25 to 50 percent dependency on nuclear electricity, and then add a Three Mile Island that goes to total meltdown, destroying a city or a land mass the size of Pennsylvania. Would the people of the United States then stand for continued operation of the plants in their own neighborhoods? Would world opinion allow further atomic operations in the face of tens of thousands of deaths from radioactivity? And if not, whence would come the electricity which we derive from nuclear sources?

No other power source carries with it so great a potential for continental blackouts. To gamble so much of our energy supply—not to mention our capital resources—on such a shaky technology is to take perhaps the most foolish risk of all.

After thirty years of operations, the atomic industry can no longer make a case as a cheap power source. If anything, it is a threat to our economic future. As economist Saunders Miller put it well before TMI: ''The conclusion that must be reached is that, from an economic standpoint alone, to rely upon nuclear fission as the primary source of our stationary energy supplies will constitute economic lunacy on a scale unparalleled in recorded history, and may lead to the economic Waterloo of the United States.''

The Fossil Alternative

Despite its horrendous economic track record, nuclear proponents argue we need atomic power for lack of other sources. The assertion is simply untrue.

For starters, current estimates of known coal reserves indicate a supply quite ample for at least another century, at costs well below those now accepted for nuclear. Coal now supplies us with half of our electricity, and, if need be, it could supply much more.

But there are serious problems connected with the use of coal. One miner dies every two working days. Overall, a coal miner is eight times more likely to die on the job than the average worker in the private sector. Every year 63,000 coal workers are disabled. Many suffer from "black lung," a crippling occupational disease against which unions have been struggling for decades.

The burning of coal also throws enormous quantities of pollutants into the atmosphere, many of them carcinogenic and some of them radioactive, as uranium and radium run through some mined coal veins. Scrubbers and other devices can drastically reduce the general pollution levels resulting from the burning of coal, but they also raise the price and don't do the whole job. According to the Solar Lobby, "Electrostatic precipitators that remove 98 percent of all particles fail to capture most of the miniscule particles that pose the greatest hazard to human health. Lead, cadmium, antimony, selenium, nickel, vanadium, zinc, cobalt, bromine, manganese, sulfate, and various organic compounds cling to these small particulates, against which evolution has provided the human respiratory system with no satisfactory defense."

Deep-shaft mining also pollutes water tables, and strip mining has become a plague. More than half of U.S. coal is now strip-mined, costing us hundreds of thousands of acres of prime farmland, devastation of our water, and leading to flooding, property damage, and death. The 1977 Strip Mine Control and Reclamation Act requires the return of strip-mined areas to something resembling their natural state, but the value of such programs is questionable.

Thus, while coal remains a far cleaner and cheaper energy source than atomic power, its costs are not to be scoffed at.

Oil and Gas

The same goes for oil. It is impossible for a private citizen to meaningfully estimate world-wide oil reserves and tell how rapidly we're running out. The 1973 Arab boycott and the 1979 Iranian revolution both led to enormous price hikes, but had little to do with reserves. In the long run, the only thing that counts in the oil business is this: *The supply and distribution of petroleum world-wide is dominated by seven multinational corporations and their partner nations. They say how much oil there is and how much it will cost.*

None of which makes oil any cleaner. Our oceans are plagued by massive spills that threaten the marine environment as perhaps nothing else in history, and from which the owners of the ill-designed, ill-managed tankers sail away unscathed. Petroleum workers suffer inordinate cancer rates because of the inhalation of toxic chemicals at the refineries. And we all suffer carcinogenic assault from the burning of gasoline in cars and other oil-fired machines just about every time we turn around.

Of all the fossil fuels, according to Barry Commoner, only natural gas seems to make any sense. The supplies, says Commoner, are abundant and the price, if controlled, could be reasonable.

There are problems with it, particularly transportation and storage. Liquified natural gas (LNG) is highly explosive and has already brought on accidents and death. LNG terminals, used largely for imported gas, have become a major environmental sore point, and are almost as unwelcome as atomic reactors.

But according to Commoner, domestic supplies are ample, and the gas can be moved relatively easily by pipelines, at least some of which can be put underground. Most important, the main ingredient in natural gas is methane. That, says Commoner, makes it "the ideal bridging fuel" to a solar economy. It is, he says, "the same fuel that can be produced from organic matter which, if we remember our basic biology, represents energy captured from the sun by photo-synthetic plants. Methane can be produced by a few simple techniques from garbage, sewage sludge, agricultural, forestry and food-processing wastes, or crops grown for the purpose."

Thus, he says, switching our energy machinery to natural gas would allow us to switch once again, easily and efficiently, to solar sources as the gas runs out.

Overall, the actual quantities of natural gas available are a matter of contention. But there remains no doubt that the supply is monopolized by multinationals and that the price has been decontrolled, sending costs through the roof.

Fusion

Also through the roof are the costs of fusion energy, an "alternative" form of nuclear power. The fusion reaction involves the merging of hydrogen atoms under conditions of extremely high temperatures. Proponents see it as a limitless source of centralized power that can be based on a limitless fuel—water.

Though fusion might well cause less radioactive pollution than fission, it would still cause some, notably in the form of tritium releases. And as a usable technology, fusion is many years and many billions of dollars away, far more a dream of the future than mass applications of solar and conservation sources. Fusion is an unproven, expensive experiment with no guarantees except for escalating research and development costs. It also represents yet another centralized, high-technology energy source that would remain well outside the control of the communities it is supposed to serve.

Throughout the entire fossil-nuclear debate, one additional fact stands out: atomic energy can do very little to cut our use of imported oil. Nuclear generators supply a single, very specific form of energy, electricity. Just 10 percent of the oil we burn goes to generate electricity. Supplanting that share would have only a marginal effect on how much oil we import.

Overall, despite environmental costs and the artificial prices imposed by the energy cartels, a mix of fossil fuels does remain available to keep the economy going.

But it would be foolish to plan long-term reliance on them. The burning of coal and oil raises the carbon dioxide level in the atmosphere, leading to a planetary ''greenhouse effect'' which is raising the temperature of the general environment and threatening ecological chaos of the first magnitude.

And though fossil reserves may hold for a longer period of time than uranium, they are still finite. Both fossil and nuclear fuel supplies are dominated by multinational corporations, in many cases by the *same* multinational corporations. Their applications are capital-intensive, creating few jobs per dollar. They fit only into a system of large-scale, centralized power generators over which individual citizens and small communities have little control.

Those generators also throw off, in the form of waste heat right at the source, between a half and two thirds of all the energy they produce. Another ten percent is lost in transmission.

The continued use of these conventional sources would drain enormous quantities of our social resources over the next few decades, not to mention the costs of cleaning up afterwards.

In the short run, coal will be cheaper than atomic power, and the supply is ample. Its health and ecological costs are staggering, but remain a whole order of magnitude lower than the nightmares imposed on us by the nuclear option.

In the long run, fossil and nuclear fuels have one final key element in common—neither is necessary.

Solar Power Now!

The technology for using solar power on a modern mass scale has been proven and available for decades. In 1952 a blue-ribbon commission reporting to Harry Truman on the state of the solar art predicted that by 1975 there could be 13 million solar-heated homes in the United States.

But in 1975 a Federal Energy Agency study reported that half the 866,000 single-family homes built that year were equipped not with solar power, but with electric heat, the most costly and inefficient source on the market.

The FEA report also pointed out that given 1975 electrical costs, every one of those houses could be more cheaply heated with solar features.

What happened in those twenty-three years between the Truman predictions and 1975?

One thing was Dwight D. Eisenhower's "Atoms for Peace" speech, which held out the promise of cheap, infinite nuclear energy. Hidden in that message were some political decisions of enormous impact.

For nuclear power held (and holds) one inescapable attraction: it is a monopoly product.

Atomic reactors require huge sums of capital to build. Their basic operations can be mastered only by a technocratic elite. And they make sense only in a system of extreme centralization. Their capital and organizational structures are beyond the control of the individual or small community. They fit only into a society with a highly centralized vision of political and economic power, which the United States of the 1950s certainly was.

An integral part of the vision was the electric utility industry. Essentially the invention of Thomas A. Edison and a business promoter named Samuel Insull, the early utilities built their capital base by selling electricity for community services such as street lighting and public transit.

But private factories found they could build their own on-site power plants and supply themselves with juice a lot cheaper than it could be gotten from distant utility generators. In fact, because so much power was lost in transmission (at least 10 percent), the factories could supply not only themselves but their near neighbors and still beat the utilities.

For Insull and Edison, that would have meant the end of the game. Their business depended on one thing—no competition.

So they charged the factories less than it actually cost to produce and ship the juice. Their prices were just cheap enough to guarantee that the factories wouldn't generate their own power. Then they built their profit margins on the municipal customers and individual home owners who had nowhere else to go.

Utilities further persuaded the legislatures to give them special status as private monopolies (the only ones in the United States) with guaranteed immunity from bankruptcy. They also got a booster known as the "rate base," allowing them to automatically base their profits on how much they spent to build generators. The more they spent, the more they could make.

With a noncompetitive rate schedule, with a profit system designed to encourage capital expeditures, and with public guarantees against bankruptcy, the utility system was custom-made for the introduction of massive power plants, including atomic reactors.

Indeed, the whole American energy grid was geared for both centralization and spiraling consumption. With an apparently infinite supply of fossil fuels and with atomic power holding the promise of still more, there seemed no reason to question the limitless expansion of the network, or the propensity of American industry and consumers to waste energy with little care for efficiency or the future. Between 1950 and 1979 the American population increased 45 percent, energy use 250 percent, and electrical consumption 600 percent. By the mid-seventies the United States, representing 6 percent of the world's population, was consuming more than a third of its energy resources.

The "Soft Path"

But the 1973 oil embargo and soaring prices that followed it changed all that. Demand curves slipped, undercutting the projections on which utilities had based their nuclear expansion programs. The rise in oil prices forced a rise in the cost of building reactors. The one hundred-plus orders placed in the early seventies dropped to a mere trickle.

In the fall of 1976, Amory Lovins, a young physicist and researcher for Friends of the Earth, published in *Foreign Affairs* a seminal tract on the energy dilemma. There were, he said, two ways our future could develop. One was the "hard path" of increased fossil and nuclear consumption, involving a continued reliance on complex, centralized, large-scale generators.

The other was the "soft path" of simple, small-scale, decentralized technologies, powered by renewable fuels and made possible through recycling and increased energy efficiency.

Lovins's thesis challenged energy policy to its core. With energy being generated at the level of the home, neighborhood, and factory, the need for centralized power might disappear entirely. "We really need no big generating plants of any kind," he told me in a 1977 telephone interview. "We could be running the country with no central power stations. Electricity costs twice as much in many cases to deliver it as it does to generate, and pretty soon people are going to realize that generating power with nuclear reactors is like cutting butter with a chain saw."

The publication of Lovins's article dramatically changed the movement against nuclear power. Printed just after the first two occupations at Seabrook, Lovins's thesis finally offered the antireactor movement an alternative.

For most nuclear opponents, atomic power was, in and of itself, a profound evil, worth stopping at almost any cost. But doing that in the real world meant producing the power some other way.

The "soft path" thesis posited an energy scenario in which less over-all energy—and especially less electricity—would be required. And one in which more—and eventually all—of that energy would be produced with nonpolluting, renewable resources that could not be monopolized by major corporations.

First and foremost, the scheme, as it became popularly interpreted, focused on increased energy efficiency—conservation. Fossil and nuclear generators waste half to two thirds of their heat right at the source, even before transmission losses are taken into account. Massive urban high-rises come with windows that won't open, lights that don't turn off, and offices that must be air-conditioned in winter. Millions of homes and factories have been built without insulation, with loose-fitting doors and windows, and with wasteful electric heating systems. Transportation remains centered on the automobile and subsidized highways while the rail system, which is far more energy efficient, has been left to deteriorate.

According to Jimmy Carter (among others), one half the American energy supply could be saved outright merely by insulating and redesigning buildings, improving mass transit, and tightening up our heating systems and machinery, all without a significant interruption in the way people live, except to cut down on unemployment. Per capita, West Germany and Sweden use half the energy of the United States, yet enjoy a comparable standard of living.

Richard G. Stein, in *Architecture and Energy*, estimates that spending $100 each to upgrade 40 million oil burners in U.S. homes— a $4 billion investment—would save us $2 billion per year in oil consumption, providing tremendous employment, while hardly affecting anyone's life-style.

There is more than enough fat in the American energy budget to cushion a wide range of simple basic technological adjustments with little impact on the economy other than to save energy and create jobs.

Recycling

As late as the mid-1960s most bottles blown in the United States were reusable, and deposits assured returns from, among others, an army of elementary and junior high school students who picked them up for spending money. But by 1970 many bottle and can conglomerates had decided to cut labor costs by shifting to no-deposit containers. The result was several million tons of expensive materials out the window along with, according to EPA figures, some 22 trillion BTUs of energy.

By the mid-seventies "bottle bills" began popping onto state ballots around the country. They prompted an industry campaign parallel to those against the nuclear referenda. According to beer magnate William Coors (one of the nation's leading anti-union manufacturers) by 1978 the bottling industry spent $20 million to fight various bottle bills. In Dade County, Florida, alone they outspent the recyclers $180,000 to $1,742.

The bottlers argued the bills would cost money and jobs and wouldn't clean up the highways. But they ignored energy. Recycling the glass, steel, and aluminum in cans and bottles would save the fuel used to mine and mill them. According to the Environmental Action Foundation, a national bottle bill would save the equivalent of 81,000 barrels of oil each day, enough to produce the electricity used by New York City and Chicago combined. According to the EPA, such a bill might also wipe out 10,000 existing jobs in the bottle and can industry. But it would replace them with 80,000 to 100,000 new jobs in the distribution and retail trades.

By the spring of 1978, five states had passed bottle bills and others were getting closer.

Furthermore, because most of the energy used in producing primary metals goes to extracting them from ores, recycling them can save enormous quantities of fuel—87 percent for copper, 96 percent for aluminum, 63 percent for lead, 63 percent for zinc. The savings extend to other products, such as paper, where 70 percent of the original energy can be conserved through recycling.

Co-Generation

An integral part of recycling and increased energy efficiency is a method of shared power production widely used in Europe—co-generation. The system hinges on using waste heat from power generators to heat or cool buildings. It is a simple technology and is spreading fast. According to Commoner, small-scale Italian electric generators, now available at $400 per kilowatt (well under nuclear costs) can also heat and cool the buildings they light. The *Wall Street Journal* says co-generation in general can save 10 to 30 percent of our over-all energy costs.

It can also save its users from centralized blackouts. One 5881-unit apartment complex at Starrett City in Brooklyn, New York, claims energy savings from 1974 to 1978 of $5 million through the use of a co-generating system. Because of its "energy independence," it was not affected by the devastating power blackout in 1977. "We were lit up like a Christmas tree," says building manager Robert Rosenberg.

Sun Rays

The solar heating and cooling of homes was known to the ancient Greeks. "We should build the south side [of our houses] loftier to get the winter sun, and the north side lower to keep out the cold winds," wrote Xenophon in *Memorabilia Socratia*. Similarly, today we should also build big windows on the south side, smaller ones on the north, and install movable insulation to cover all of them at night, thus cutting down on heat loss. Hardwood trees, offering shade in summer and letting sun through in winter, should be planted to the south; evergreen trees, which block the cold winds in winter, should be planted to the north. High-mass concrete, masonry, and water storage systems can be built in for long-term storage of direct solar heat.

Modern solar panels and design features can now provide 50 to 90 percent of all heating and cooling needs—in North and South—with additional new home-building costs ranging from zero to $6000. Old homes can be retrofitted at greater cost, but as the price of fuel soars, the payback period gets shorter and shorter.

Meanwhile, a Federal Energy Agency Task Force on Solar Commercialization has shown that a $450 million government investment in photovoltaic cells could yield, by 1985, mass production of solar-generated electricity for just $500 per kilowatt—half the current cost of nuclear. The program would use the new technology only where it is already cost-effective, thus making the cells pay for themselves as their development progresses.

Photovoltaics are thin silicon or chemical wafers which can convert sunlight directly into electrical current. As a small-scale, nonpolluting technology, it would be ideal for making homes, offices, and factories energy independent. "We have the perfect area for installing photovoltaics everywhere," says John Abrams of the Massachusetts Energy Resource Group. "You can put them on the rooftop of a house and just run the electricity right inside. If you need more, they can be set up on a garage or in a field."

The Wind

Back in the 1930s, some six million windmills dotted farms throughout the American countryside. Some of the better models, such as those made by the legendary Jacobs Brothers, are still operable, and fetch prices in excess of what they sold for forty years ago.

Wind power is now cost-competitive with nuclear for electricity. According to a study by the U.S. Department of the Interior, the power equivalent of ten large central generators could be supplied by a wind-powered pumped-storage system in the Pacific Northwest. Medium-sized community generators are now working at Block Island, Rhode Island; Cuttyhunk, Massachusetts; Clayton, New Mexico, and elsewhere. The world's largest windmill now operates at Twind, Denmark; it was built by college students and faculty at a cost of $300 per kilowatt. U.S. Wind Power of Burlington, Massachusetts, is now marketing wind generators at $500 per kilowatt (half the cost of nuclear) and is backed up with orders. California has ordered twenty of them for use in the state's water system. Overall, the field is just beginning to show its limitless potential.

Wood

Wood energy is also experiencing a real boom, especially in New England. The manufacture and sale of clean, efficient wood stoves has become one of the leading growth industries in the Northeast. Firewood usage in New England alone has soared by a factor of six since 1970, with nearly one fifth the homes in Maine, New Hampshire, and Vermont relying on wood for their primary source of heat. At least 150 of the region's industrial firms have converted to wood, as have several municipal electric companies.

Wood can also be used for base-load electrical generating. A 50-megawatt electrical generator at Burlington, Vermont, will be fueled with wood refuse from nearby lumber mills. Its waste heat will be recycled through large-scale commercial greenhouses surrounding the plant.

The New England Regional Commission, a business group, has listed New England forests as representing the energy equivalent of three billion barrels of oil.

All of it renewable. Most of America's woodlands have been logged—some of them twice—and are in many places in poor condition, in desperate need of thinning and maintenance. With proper care, those forests could provide both energy and thousands of accesible, productive jobs. If it's done right, we should also come out of it with healthier forests and a tremendous boon to the industrial economy.

Biomass

There are, using the concept of biomass, other energy "crops" available. Technology has advanced rapidly for converting agricultural crops such as corn, corn stalks, grain, and even sea kelp into alcohol and methane for industrial and transportation uses. Brazil has now begun the mass cultivation of sugar cane and manioc root crops for the sole pupose of producing alcohol, which is then to be mixed with gasoline for "gasohol," a fuel capable of powering the nation's automobiles. The Brazilian program is scheduled to replace 20 percent of that nation's gasoline consumption with the produce of less than 2 percent of its farmland. By the year 2000 the Brazilians hope to use the home-grown alcohol to phase out all oil imports.

Rural and urban garbage can also be fermented into alcohol, or processed for the accumulation of methane gas. Farm manure and urban sludge can, through proper composting, generate huge quantities of clean-burning fuel.

Trash can also be burned outright, as is being done in St. Louis and elsewhere. The process produces energy while reducing environmental and financial costs from landfill dumps. It also encourages recycling through the separation process required to isolate the burnables.

Geothermal

Still another natural source of energy is the heat of the earth's core, geothermal energy. For centuries humans have tapped this source, using it even to heat entire towns. In recent years use of the process has speeded up. In some cases, however, local residents have complained of heat and chemical pollution, and others question its long-range ecological impact. Though a promising source of power, geothermal technology and its by-products may require more extensive study.

Hydro

Still more power is available from our rivers and oceans. An ocean tidal generator has been in operation at Rennes, France, for many years, and others are being planned the world over.

Also on the drawing boards is a technology called Ocean Thermal Energy Conversion (OTEC) aimed at tapping the temperature differentials of different levels of the ocean.

Power is more immediately available from one of our oldest sources—hydroelectric dams. Though many American waterways are dammed, some sites for small new facilities are still usable. In addition, the Army Corps of Engineers has estimated that simply expanding or rehabilitating existing dams could yield an additional capacity equal to more than twenty-five 1000-megawatt nuclear power plants.

Unfortunately, small-scale hydro has been attacked by atomic backers in at least two instances. In Springfield, Vermont, local townspeople and their government have been retooling and modernizing the town's hydroelectric facilities, forming a municipal utility

ing for local energy self-sufficiency. They have been
the way by the Central Vermont Power Company, a part
Vermont Yankee.

v Hampshire, where at least a score of hydroelectric facilities
ding idle, state workers dynamited the 800-kilowatt Lochmere
Tilton. Critics charged that this was tantamount to sabotage to
help guarantee a rising electrical demand that would justify Seabrook
Station. According to James B. Walker, a local businessman, the
Meldrim Thomson administration had told the town that state workers
would be working on a canal next to the dam. But, says Walker,
"When work started, one of the first things done was to smash out the
turbines and dynamite the inlet structure in the powerhouse." As
president of the local conservation society, Walker wrote the Statehouse charging that "there is a formerly large, but presently untapped
reservoir of energy in New Hampshire," and wondered what could be
the "rationale behind the apparent ongoing action in abandoning
hydropower."

The Solar Threat

The fight in Springfield and the dynamiting of the Lochmere dam
underscored the "threat" of solar power.

Taken as a whole, increased energy efficiency, recycling, co-
generation, and a drastic increase in the use of renewable sources
could revolutionize the economics of American energy.

According to the Solar Lobby, by the year 2000, with an investment
of roughly $2.5 billion per year—the bottom line cost of a new nuclear
power plant—we could be getting 25 percent of our energy from the
sun. That's six times what we now get from nuclear. By the middle of
the next century, we could be using natural sources entirely. The
technology is there. The problem is politics. "There's no doubt about
what we can do," says Amory Lovins. "But it may take several years
to clear away the institutional barriers."

That last may be the understatement of the century. The utilities
industry and the energy cartels live in mortal, active fear of a solarized
economy. They are in the business of selling energy; the sun gives it
away for free. By making power available in forms that homeowners,
tenants, and small communities can control on their own, a deluge of
low-cost windmills, solar collectors, wood stoves, small dams, and
efficiency improvements could destroy both the utility industry and

the energy cartels. Solar would substitute human labor and neighborhood initiative for high-priced machinery and centralized, entrenched bureaucracies. If families, neighborhoods, and towns could generate their own heat and light, who would need the energy monopolies, their power lines, or their nukes?

In the wake of Three Mile Island, opinion polls have shown an overwhelming public desire to do whatever is necessary to bring on solar power. Energy from the sun is like apple pie and motherhood: you can't really be against it.

You can, however, say it's "impractical." To combat the "menace" of a solar society, the energy cartels have employed some time-honored techniques:

• *Advertising*: Millions have been spent by the energy cartels buying media time to tell us that solar power isn't realistic, a dream of the future. Those lies are being paid for with our money, which we sent in with our gas and electric bills.

• *Misdirection*: Government solar research money—that is, *our* money again—has been gobbled up by corporations with dubious motives. This funding has gone into investigating basic solar technologies that have been known to work for years, if not decades. By lavishing thousands of dollars on unnecessary, marginal experiments, government and industry have continued to promote a public image of solar as being an expensive, fringe technology. They've also managed to keep federal money from going into development and production of solar hardware, where it's really needed.

• *Perversion*: Solar power can also mean many things to many people. To Peter Glaser of Arthur D. Little, the Cambridge think tank, it means gigantic orbiting solar satellites designed to beam back 5000 megawatts each in the form of microwave energy. Glaser's space stations would be seven miles long and two and a half miles wide and would cost about $8 billion each. One plan calls for sixty such collectors, with research and development funds estimated at around $50 billion.

Other, more earthly schemes involve centralized square miles of mirrors beaming rays at a "power tower" which would then ship megawatts around the country (see next article). While this scheme could be useful in some cases, it's also somewhat beside the point. We already have many square miles of space waiting for solar collectors—on the walls and roofs of buildings already standing.

• *Outright Attack*: Meanwhile, numerous utilities are actively discouraging homeowners from filling those roofs and walls with solar features. Your bill may go up, for example, if you replace your electric hot water heater with a solar one, as many utilities systematically raise rates for those who decrease their consumption. Their claim is that if you cut back on what you use but still remain plugged into the grid for lighting and so on, then the utility must continue to provide the backup capacity, and is thus justified in charging you more for using less. One windmill owner in Leverett, Massachusetts, should therefore not have been surprised when she discovered a meter man routinely inspecting her house to find out why she was using less juice. With such logic and enforcement, it's not hard to imagine a "solar tax" in the near future.

In a real sense we've all been subjected to a "conservation tax" since 1973. Since the Arab embargo, electrical consumption growth rates have plummeted. In many instances, households and communities have lowered their consumption in absolute terms.

Yet whenever that happens, utilities complain the drop in energy cuts their profits, and therefore they should be granted rate increases. Thus people who use less often pay more with no real incentive to conserve or go solar.

• *Co-optation*: Meanwhile, the energy cartels themselves are quietly wading into the solar field in a serious way, for their own purposes. According to the Center for Science in the Public Interest, large corporations took out thirty of the forty-seven solar heating patents granted between 1960 and 1977. Some of the "smaller" companies, such as Grumman Aircraft, actually seem to be interested in building and marketing solar hardware. But General Electric, Westinghouse, and the oil companies are more likely to bury the patents as soon as they get their hands on them.

• *Monopolization*: Should solar technology become unstoppable, however, the multinationals will be right in there selling it. All those buried patents will see daylight real fast if it becomes evident that the flood of solar sentiment is not to be denied. And with them would undoubtedly come legislation and commercial promotion designed to take the big energy corporations as far toward dominating the solar field as they can go.

Programs are already underway to allow utilities into the insulation business, making it possible for homeowners to charge their costs off on their electric and gas bills. It seems sensible and a reasonable method of financing increased energy efficiency. But the utilities have been at the root of our conservation problems, and their progress into the field must be viewed with suspicion. A logical further step would be the "leasing" of solar hardware, which would only leave in the hands of the energy bureaucrats what should be a decentralized, community industry.

Solar Power Is People Power

The beauty—and liberating power—of solar energy is that there is no economic advantage to centralizing it. It demands small-scale applications that individual families and communities can best engineer.

As such, it is the ultimate threat to the utilities industry and to the energy multinationals that are the world's most powerful corporations. They have a clear and obvious interest in stopping the spread of solar power, and they seem to be acting on it. "The suspicion is almost unavoidable," says U.S. Senator Gaylord Nelson, "that the giant firms, because of their large investments in nuclear technology, hope that solar energy will not gain rapidly."

Adds windmill expert William Heronemous: "I have no doubts that the major utilities have played a role in sabotaging natural energy development. They certainly have a lot to lose."

Can we do it? Can we rebuild our industrial system around natural energy sources? Can we become a solarized society?

The answers are political, not technological. The barrier is not science but money, money to finance mass production of solar hardware, money to be made available to homeowners, tenants, and communities for the purchase and installation of solar features.

That money can and should be redirected from the utilities, reactor producers, and oil cartels that have been ripping us off for so long, and for so many billions. Those dollars represent resources that belong to all of us. They can and should be available for the betterment of society as a whole, rather than continuing to pour into the overflowing coffers of Exxon, General Electric, and the local utility.

World opinion now seems to be swinging hard and fast toward the demand for rapid development of a solar future. It's time we grabbed that option with both hands. The real question is not whether we can go solar, but "What choice do we have?" To continue on the fossil-nuclear path is to continue down a road of blackouts, shortages, soaring prices, unemployment, debilitating capital loss, ecological catastrophe, and radioactive suicide. Continued reliance on large-scale conventional generators is, literally, a dead end. We must find something else, and that something else is obviously solar power. "We decided to put a man on the moon in ten years," says Dr. Erich Farber, a Florida solar pioneer. "We didn't have the slightest idea how to do it then, and we did it in nine years. We already have the answers on how to use solar energy, and if we would make the decision to do it, we could do it much better, much quicker, much easier.

"When people ask me how soon will solar energy become widespread, I always tell them, that's up to you."

☆☆☆

While most solar advocates see renewable technologies as naturally decentralized, there is no guarantee they'll stay that way. When utility, oil, and nuclear advocates speak of solar power, they talk in terms of giant collectors and outer space. There is no law, natural or otherwise, that says solar energy must be democratically owned or community controlled.

A concrete example opened for business at the Sandia Laboratories outside Albuquerque in the fall of 1978:

Government's Leap into Solar: New Mexico's Power Tower Ushers in Solar Electric Age
(Pacific News Service, October 31, 1978)

Albuquerque

IN THE MIDST OF THE BLAZING NEW MEXICO DESERT there stands a huge gray monolith, two hundred feet high, flanked on its north by 222 expensive "heliostat" mirror systems.

On October 27, 1978, Sandia Laboratories technician Debby Risvold aimed one of those mirrors at a boiler perched atop the tower. In so doing, she inaugurated a new era of solar-energy research.

But whether the path of those rays marked a direction in which solar research should, in fact, go has become a matter of debate, one whose resolution will say much about our energy future.

The "monolith" is officially called the Solar Thermal Test Facility (STTF), but is commonly known as the "power tower." Under construction since July, 1976, the $21 million facility is at the core of a major Department of Energy push towards centralized solar power.

Inside the tower sits a one hundred-ton elevator, capable of lifting that much again.

Outside, the 222 movable heliostats—each having twenty-five highly polished movable mirrors mounted on it—are designed to focus the rays of the powerful desert sun onto narrow openings spaced up the tower. In May, 1977, Sandia engineers focussed 71 heliostats (1775 mirrors) at the tower's 114-foot mark. According to an official press release, "the concentrated sunlight produced temperatures of over 3000 degrees Farenheit, melting a two-foot by three-foot hole in [a] quarter-inch-thick steel plate in less than two minutes."

The goal for the current testing is less spectacular. This time the target of the focused sunlight was a test boiler rated at one megawatt, built by Boeing. Sandia technicians intend to monitor the boiler's performance at a $5 million computer center about a quarter mile from the tower.

A subsidiary of Western Electric, Sandia is better known for its work in the nuclear field than for its ventures in solar energy. Its headquarters and test facilities, including the one hundred-acre STTF site, are within the confines of Albuquerque's huge Kirkland Air Force Base. The main office sits two blocks from the Air Force's "Atomic Museum," where replicas of "Little Boy" and "Fat Man," the Hiroshima and Nagasaki bombs, are mounted amidst a veritable pageant of nuclear clippings, memorabilia, and scale-model weaponry.

Sandia itself is a private corporation. As Bob Gall, a company public relations man, explains it, Sandia manages STTF operations for the Department of Energy "without profit or fee. We're going to be testing a wide range of mirror arrays, heliostats, and boilers. The results we get here will help determine the best features for future applications of this kind of technology."

"This kind of technology" refers to large-scale central-generated electricity from the sun. The power tower itself will generate no juice, serving instead as a testing ground for prototypes. The Boeing boiler is scheduled to be followed by a 5-megawatt model from McDonnell Douglas. Other familiar names such as Martin-Marietta and Honeywell have designed prototype heliostats and mirror arrays.

"The facility could generate 6.5 megawatts worth of heat," explains project manager David Darsey. "We may pick a time to optimize power, but mostly we'll be running at 5 megawatts and putting the different equipment and alignments through various tests."

The focused sunlight will be aimed at five openings spaced twenty feet apart from 120 feet up the shaft to the top. The facility, which has consumed some 4500 cubic yards of concrete, now employs twenty workers.

Darsey's computer control center is a maze of consoles, modular computer units, and wall charts. The building is linked to the tower by a Kafkaesque tunnel that carries a half-dozen large metal electrical conduits under the mirror arrays. Some of the wiring, Darsey explains, is for manipulation of the tower equipment, while other hookups are for monitoring devices which will check the performance of boilers riding up and down the shaft in that huge elevator.

The Sandia tower is a key link in what official energy planners see as a potentially profitable way to grab the sun.

For although the STTF boilers won't be hooked up to any generating equipment, the tower will serve as a prototype for a 10-megawatt plant being built at Barstow, California, by the DOE in partnership with Southern California Edison. DOE timelines call for two such 10-megawatt "pilot plants" to enter the grid by the early eighties, with at least one "demonstration plant" in the 50- to 100-megawatt range to come on line by 1985, and with 100- to 300-megawatt "commercial plants" proliferating soon thereafter.

Sandia developers argue the power tower is safe and pollution-free. "We don't think there's a real chance of anyone getting hurt here," says project manager Darsey. "But we're not taking any chances. Nobody is going to be allowed to walk in front of the mirrors while the sun is shining.

"And if anything really does go seriously wrong, the worst that could happen here is that we'll lose our power."

Such precautions seem almost humorous when considered alongside the potential hazards of nuclear and fossil facilities.

But serious questions have nonetheless been raised about the STTF and projects like it. "I find it offensive that a laboratory building plutonium triggers is also moving into solar," says Richard Grossman of the Washington-based Environmentalists for Full Employment.

Indeed, government energy planners have displayed a remarkable knack for "pioneering" solar power in a military-industrial context. Another key link in the official solar program is the windmill test facility at Rocky Flats, Colorado, operated by Rockwell International. Centering on some three-dozen test towers and windmills, the facility feeds off stiff breezes that then flow over the Rocky Flats plutonium mill just a mile downwind—also managed by Rockwell International. The defect-plagued 100-kilowatt windmill at Plum Brook, Ohio, has been under the auspices of NASA, and sits within a mile of an inactive nuclear reactor.

The presence of large military-related corporations on the receiving end of government solar funding galls many researchers who feel the best work has been done by small independents, particularly in areas like New Mexico, where predominating sunshine makes the climate ideal for decentralized units. "We have the highest per capita number of solar houses in the whole country," says Keith Haggard of the prestigious New Mexico Solar Energy Association. "This state is perfect for small-scale, passive systems and it really is pioneering the field."

"It really is distressing," adds Albuquerque solar activist Dede Feldman, "to see them spending $21 million on that project when so many people with small units could use the money. It makes you wonder what they're driving at."

At best, the power tower evokes a certain ambivalence among most solar proponents. "To the extent that you want to repower existing plants in the West, the power tower might not be a bad idea," says Denis Hayes of the Worldwatch Institute and a moving force behind Sun Day. "To the extent they keep the size down, there doesn't seem to be much indication of environmental damage. But economically, I'd bet on photovoltaics."

According to Hayes, fiscal 1979 will be the first year that the federal photovoltaics budget will outstrip that for power towers. "The tower's applications are limited to the Southwest," he says. "It is also limited to situations where you need electricity."

Beyond that, he adds, "indications are pretty strong that the optimal scale for the concept may well be under one hundred megawatts. Probably the program is getting more [money] than it deserves. But time will tell that best."

Meanwhile, the issue of large-scale centralized solar facilities remains a major unresolved question in the energy debate. It's no secret that when James Schlesinger thinks about solar power, he thinks not in terms of small-scale solutions, but about multibillion-dollar space stations and a centralized corporate-military program.

What role Sandia's power tower plays in all that is not yet clear.

But its opening this fall does magnify the urgency of delineating who will develop and own our renewable energy resources. It raises serious questions about how far the government intends to go in centralizing solar energy.

And it can't help but make one wonder about how much of the sun the Department of Energy intends to hand over to the war industry.

☆☆☆

About as far from the power tower as one can get on the solar side is the New Alchemy Institute in Woods Hole, Massachusetts. Firmly committed to decentralized self-sufficiency, the New Alchemists have pioneered simplified, low-cost solar living with the utmost in imagination and resourcefulness, a sort of new-age Yankee ingenuity.

Along with hundreds of other pioneers around the country, they are laying the foundations for an economy based on renewable energy and cyclical, self-regenerating food production, probably the only kind of economy that can guarantee human survival into the next century.

New Age Farmers Launch Ark
(*Politicks*, January 3, 1978)

Woods Hole, Mass.

"THE SUN COMES IN and strikes the algae," explains New Alchemy coordinator, Ron Zweig. "Algae cells absorb radiant energy and heat the pond. They also oxygenate the water through photosynthesis and purify it by metabolizing the fish wastes. The fish, of course, feed off the algae."

Then the humans feed off the fish. And if they're smart, they also heat their buildings with the solar energy absorbed by the algae into the ponds.

Five feet high and wide, the "ponds" are made of thin translucent sheets of fiberglass fused into thick watertight cylinders to create what may be the world's most efficient combined solar heating and food production units. But what makes the system most remarkable is that it is no theoretical model—it's being tested and used by a shoestring collective of practical ecologists at the New Alchemy Institute here at the southwestern corner of Cape Cod.

Last month, the New Alchemists marked the end of their fifth season with a fish harvest and feast, followed by a day of seminars and guided tours of their twelve-acre spread. The space includes two dozen fish ponds, several domes and windmills, a worm ranch, trenches for growing insect larvae, a "circular river," a large organic garden, and a solar greenhouse known as "the Cape Cod Ark."

"In the early seventies," recalls John Todd, a founding member who holds a Ph.D. in biology, "we were a group of individuals with a wide range of degrees and backgrounds, and not one of us knew how to make a single piece of the world work. We knew that science was anything but neutral and that it had to be restructured—given a human dimension. But we also knew this had to be done in a practical way."

Todd and a few others who had come to know each other at educational centers in Ann Arbor, San Diego, and around the Oceanographic Institute at Woods Hole, finally decided to try their hand at building food- and energy-support systems on a piece of leased land ten miles from the ocean. "You can't separate the way society sustains and feeds itself from its psychology and its politics," says Todd. "We were looking at a country bombing the hell out of Vietnam, while at the same time feeding itself with corporately owned, petroleum-dependent agriculture. We came to the conclusion that there was no separating those two conditions. So the question we put to ourselves was: "How do you involve people in raising food in a way that is economic and attractive?"

To chronicle their answers, the New Alchemists have printed four substantial journals, material from which has been edited into a full-length book just published (*The Book of the New Alchemists*; E.P. Dutton, $6.95).

On Saturdays from May through October they've also hosted several thousand visitors. The last group of 1977 was told by Nancy Todd, another co-founder, that "we started, basically, when we saw that people were moving back to the land, but were failing for lack of knowledge and tools. What we've tried to do here is find ways of providing basic support as cheaply and simply as possible."

"We're basically aimed at the urban situation," says Zweig, who oversees the Institute's ponds. "What we're doing is trying to perfect a system where people can raise protein in a very limited physical area, and at the same time make use of solar heat. The process is simple, and there doesn't seem to be any reason why it can't work."

Indeed, the core of the system is nothing more than a series of translucent tubes ranging in cost from $65-$250, depending on quality and whether you assemble them yourself. The tubes are filled with water, green algae, and tilapia, a genus of fish from east Africa which are, as Zweig explains, "hardy, fast-growing, essentially vegetarian, disease resistant, and real good to eat." They can thrive indoors and to a large extent live off vegetables and cuttings from interior plants. They are considered tropical and, should a few escape, could not survive northern winters to wreak havoc with the ecology, a precondition the New Alchemists set for all their imported plants and animals.

"Last year," says Zweig, "we got ten times the return-per-acre that anyone has ever recorded for fish, including the Chinese." About thirty pounds of this year's crop made its way onto the table of a season's end communal fish fry which testified to the tilapia's taste.

The New Alchemists have also designed something called a circular river, where water is pumped—by windmills—out of one of the in-ground tilapia ponds, through a series of pools which serve as natural filters and feeders, and then back into the pond. "Biologically," says John Todd, "the fish believe and react as if they're living in a large lake."

The New Alchemists also operate a "midge works," where insect larvae are grown on burlap sheets dipped in shallow trenches for the feeding of special breeds kept in a back-lot pond. A large worm-farm composting operation produces additional fish food, as well as helping to feed a carefully monitored organic garden which is mulched and fertilized with a wide variety of at-hand substances ranging from seaweed to maple leaves.

But the Institute's centerpiece is the "Cape Cod Ark," a ninety-foot-long greenhouse whose fiberglass panels extend twenty feet into the air and cover a series of fishponds and gardens designed to survive the bitterest New England winter. "We had to fire the wood stove maybe two weeks last snow season," says Zweig. "Otherwise, it was all direct solar energy." The Ark carries such exotic specialties as banana, fig, papaya, and grape crops, and keeps the Alchemists in fresh vegetables through the winter. Built at a cost of around $40,000, the Ark is designed, says Zweig, "to make it possible for people to go into farming without laying out a quarter-million dollars. The Ark can provide food and a modest income for small families with limited capital and land."

A much larger Ark now functions on Prince Edward Island, Canada, with support from the Canadian Government. A small group of New Alchemists operates, without land, in Santa Cruz, California, while a new Institute is under way in Costa Rica. "The idea was to spread the word and develop systems that would be applicable in many different parts of the world," says Christina Rawley, the Institute's educational coordinator. "If we can develop systems that can put people—no matter where they are—back in touch with the planet and the way it works, and do it so it pays, we'll have gone a long way toward solving a lot of problems."

The Institute has been largely supported by foundation grants and memberships which range from $10 per year to $1000. Anyone willing to make the trek to Cape Cod in the winter might call (617) 563-2655 or write P.O. Box 432, Woods Hole, Mass. 02543 for copies of the book, journals, or further information.

But for the next tilapia fish fry—solar energy, windmills, and circular rivers notwithstanding—you'll just have to wait till the snow melts.

11

Beyond Three Mile Island

☆☆☆

JUST AFTER THE INCIDENT AT THREE MILE ISLAND an unnamed United States Senator told *Newsweek* that the accident was the Tet Offensive of the antinuclear campaign. The analogy is worth exploring.

In October, 1967, a hundred thousand people rallied at the reflecting pool between the Washington Monument and the Lincoln Memorial to protest the war in Vietnam. Thousands of them then marched to the steps of the Pentagon, where they occupied the war machine's front stoop for a day and a night.

That march marked the culmination of hundreds of smaller demonstrations and years of organizing. The biggest antiwar gathering to that time, it was dramatic proof of the surging public anger over the American intrusion into Southeast Asia.

That winter, during the Tet holiday, National Liberation Front fighters poured into the cities of Vietnam in an offensive that defied U.S. military firepower. Despite the Pentagon's best efforts to convince us otherwise, it seemed clear that the Vietnamese countryside could not be controlled by military force. It was a humiliating experience for the American military elite, and it seemed to prove once and for all that there would be no "victory" in Vietnam.

Around that time, Eugene McCarthy humbled Lyndon Johnson in the New Hampshire presidential primary. Robert Kennedy then jumped into the race, and this seemed to guarantee that the Democrats would nominate a candidate committed to withdrawing from Vietnam. On April Fool's Day, Lyndon Johnson confirmed the inevitable by pulling out of the race.

But somehow it didn't work out that way. By August, Kennedy and Martin Luther King, Jr., were dead, and the hopes of a generation seemed to die with them. By November, partly on the strength of his "secret plan" to end the war, Richard Nixon had been elected President. Somewhere along the way, the polls began to show that more than half the people in the country were ready to pull out of Southeast Asia.

But it would be 1975—seven long years—before the last troops actually cleared out.

During and immediately after Tet, military officials downplayed the importance of the attack. It was, they said, a "desperate" gesture, a "pyrrhic victory" that only showed the "weakness" of the National Liberation Front. In the midst of an astonishing debacle, the Pentagon stonewalled it, resolutely refusing to admit the obvious.

We saw the same show during Three Mile Island. "We didn't injure anybody, we didn't seriously contaminate anybody, and we certainly didn't kill anybody," said Metropolitan Edison's John Herbein. The accident, added James R. Schlesinger, a Vietnam veteran in his own right, "underscores how safe nuclear energy has been in the past." For the industry, the accident was just a "big, bad publicity blast" that would eventually pass and, in the view of some, might even strengthen the business by showing how well a "near miss" could be handled.

Indeed, with seventy reactors online, ninety in the works, and another two or three dozen in the advanced planning stages, the legacy of atomic power is alive and well. Though it has become fashionable to term the nuclear industry "dead," the obituaries are premature. Killing atomic power will take far more than Three Mile Island and dismal prospects for future growth of the industry. Since the industry embodies thirty-five years of the best hopes of the world's technological elite, and the economic stakes dwarf those of any jungle war, there is every reason to believe that without a full-scale working alternative, the atomic industry could assume a low profile for a few years and return ready to grow and fight again.

Right now the antinuclear movement is mushrooming much as the peace movement did in the late sixties, when the media finally acknowledged the issue, when the public opinion polls quickly began to change, and when the war really began to inflame passions and move people to action.

The thousands of people now marching against atomic power, the flood of literature, the emotion, the political energy, the antinuclear rumblings in Congress—all are heady indicators that change is on the way.

But the size of the job at hand is staggering. This energy system we've inherited is fundamental to the entire American industrial machine. The centralized control and distribution of nuclear and fossil-fuel power is a cornerstone of modern finance. The corporations that dominate our energy supply are the world's largest.

It should be clear to all of us by now that there is far more to stopping atomic power than just ridding ourselves of a single technology. To be sure, the lives of this and all future generations depend on shutting down the nuclear industry. Never has human kind faced so insidious a threat to its ability to survive and reproduce. Stopping those reactors from being built and going into operation is a life-and-death matter for this and future generations.

But doing it means far more than merely substituting solar technology for nuclear. It means battling the most powerful corporations in the world. It means revolutionizing our industrial economy. It means transforming an autocratic energy system into a community-owned, democratically operated system. Changes just don't come more basic than that.

In 1976 some of us hoped we had elected a President who would help. As a candidate, Jimmy Carter had talked about atomic energy as a "last resort." That seemed to indicate the nukes would go to the back burner, and that a massive new push for solar power was on its way.

But little new money or leadership for solar development has come from the Carter camp. There've been a few helpful appointments, some positive steps here and there.

But Jimmy Carter has clearly reneged on his promise to treat nukes as a "last resort". As so many presidents before him, he said one thing on the campaign trail and did entirely another in the White House. His presidency has been downright tragic. If ever a national figure had an opportunity to spearhead a public campaign toward higher goals, it was Jimmy Carter. He could have gone straight to the people with the solar issue. The money was there, the public was ready, the crisis was crying out to be solved.

Instead he wallowed in confusion, caving into the oil companies on decontrol, saddling us with higher energy costs, continuing down the atomic road. He talked conservation, but portrayed it as a national sacrifice, rather than as an opportunity to tighten up our machinery and live better simply. He talked about a collective need to alter our energy habits, but threw the burden on those who could least afford it.

Candidate Carter had also promised to cut military spending. But as so many Presidents before him, he raised it.

Spending for war now accounts for fully half the entire federal budget, even though we are ostensibly at peace. It would take just a fraction of what we lavish on the tools of death to accelerate the solar industry rapidly.

And it would take the detonation of an even smaller fraction of the nuclear armaments we produce to wipe us all off the face of the earth.

Unfortunately we've lived with a war economy and the nuclear threat so long that most of us have become numb to it. Yet the military budget is the most inflationary of all the items on our tax bill. It cripples our ability to progress and colors every day of our lives with profound danger and waste. Until the billions and billions of dollars that go each year for war are liberated for social uses, and until the stockpiling of atomic weapons is ended, our efforts to build a sane, fully prosperous society will fail.

The question of who gets elected in 1980 pales in the face of twin evils as huge as the energy crisis and the weapons race. No mere President will end our war or energy problems. Nor is it likely they will be solved—or even dented—in the space of four years.

Those committed to fighting it out for a safe energy future—for any future at all—had best think in terms not of years, nor of presidential spans in office, but of decades. This one is going to take a lot of hard work, patience and a very solid collective sense of humor.

The power of social movements comes not in their ability to elect officials, but in their ability to move people at the grassroots. National elections tend to weaken popular movements by draining their local base and dividing people. The true test of 1980 will be reflected in how strong the prosolar movement can emerge in 1981.

The atomic issue will be decided not in the White House, but in the neighborhoods and at the nuclear sites.

And the successful adoption of nonviolent techniques on a large scale marks a major step forward in American community organizing.

Peaceful direct action is to politics what solar power is to our energy system. It demands the best of both the activist and the community. It has transformed angry confrontations into profound human experiences that have changed for the better both lawbreakers and police.

And though nonviolence—like renewable energy—demands a certain leap of faith, it offers us our best chance to organize a grassroots movement with the staying power needed to achieve a solar society.

One of the great defeats of the Vietnam era was that the peace movement failed to build or sustain the ongoing neighborhood organizations needed to win a long-term campaign for social change. As the war dragged on, the lack of a firm community base caused the movement to splinter and, in too many cases, to destroy itself in factionalism having little to do with the real issues affecting people's lives.

So far, the prosolar movement's nonviolent commitment and community roots have enabled it to avoid the polarization and alienation that derailed so much of the antiwar campaign. If it stays on course, it could prove as important a social movement as this country has ever seen.

It now seems inevitable that increasingly serious attempts will be made in the near future to physically stop reactor construction and operation—and mining, waste disposal and weapons production—through nonviolent mass action. At some point, somewhere down the road, one of those attempts will succeed.

At some point, too, the trickle of communities that have taken control of their local utility systems will become a tidal wave. There are now some three thousand consumer-owned utilities across the United States, serving some 20 percent of the nation's electric consumers. As a whole, they supply electricity far more cheaply, efficiently, and democratically than the two hundred investor-owned utilities from which most of us must buy our power.

Since 1960 some thirty towns and cities have authorized public takeovers of private utilities. But barriers are formidable. In Massena, New York, the campaign took six years—through feasibility studies, tough town elections, well-financed opposition from the local utility, and the inevitable court battles—before the town finally took charge. Other communities can expect fights at least as long.

And simple public ownership may not be enough. The Tennessee

Valley Authority, Bonneville Power Authority, and other large government-owned utilities are prime backers of nuclear power, and they are widely despised by their customers for their high rates and autocratic management. True public power demands grass-roots control, with basic decisions made democratically in the towns and neighborhoods.

Right now, the big electric companies are among the most potent opponents of an economy based on renewable energy sources. They've abused their special status as sheltered monopolies to promote energy waste and block community control. But then again, that's what they are designed to do. Their survival is at stake. Solar energy can undercut the utilities; public power decentralizes and takes them over. In the long run, both must come about.

And make no mistake about it—it will be a long run. The electric utilities are small-scale front groups compared to Exxon, Mobil, Shell, Getty, and the other energy cartels that stand to lose most by a shift to solar power. Fighting them for renewable energy is a job dwarfing in magnitude what it took, on the home front, to stop the war in Vietnam.

But we have no other options. The fossil/nuclear economy embodies a state of permanent war on the natural ecology by which we live. How we get our energy affects at the most basic level how we live. A society that is destroying the planet to get its power cannot hope to achieve inner or global peace. There's the real "energy war."

Indeed, our energy hawks are already calculating, in public, the possibility of a military attack on Mideastern and other oil-producing nations. If America's appetite for energy isn't quenched—and soon—there's every reason to expect another Vietnam, or worse. World wars have certainly been fought for less.

The faith of solar advocates is not so much in the simple devices of renewable technology. Rather it's in the idea that Nature will provide us with all we need if only we will understand, respect, and nurture her. Where the sun doesn't shine, the winds blow, the rivers flow, the crops grow and energy can be gotten in as many ways as we can devise. If we can split the atom, we can surely figure out how to ride the natural bounties of this amazing planet into a full, rich life for all of us.

A solar peace is not only a matter of survival, it's the promise of a prosperous, equitable future.

But getting there will demand a lot more Human Fire of our own.

Sources of Information on Nuclear Power:

**GENERAL INFORMATION
& REFERRALS**
Nuclear Information & Resource Service
1536 16th Street, N.W.
Washington, D.C. 20036

CIVIL LIBERTIES
Citizens Energy Project
1413 K Street, 8th Floor
Washington, D.C. 20005

ECONOMICS
Nuclear Information & Resource Service
1536 16th Street, N.W.
Washington, D.C. 20036

EVACUATION PLANS
Critical Mass Energy Project
P.O. Box 1538
Washington, D.C. 20013

JOBS
Environmentalists for Full Employment
Room 305
1101 Vermont Avenue, N.W.
Washington, D.C. 20005

LOW-LEVEL RADIATION
Bob Alvarez
Environmental Policy Center
317 Pennsylvania Avenue, S.E.
Washington, D.C. 20003

PROLIFERATION
Mobilization for Survival
1213 Race Street
Philadelphia, Pennsylvania 19107

REACTOR SAFETY
Union of Concerned Scientists
1208 Massachusetts Avenue
Cambridge, Massachusetts 02138

SITING & LICENSING
Marc Messing
Environmental Policy Center
317 Pennsylvania Avenue, S.E.
Washington, D.C. 20003

SOLAR POWER
Center for Renewable Resources
Room 1100
1028 Connecticut Avenue, N.W.
Washington, D.C. 20036

Institute for Local Self-Reliance
1717 18th Street, N.W.
Washington, D.C. 20009

URANIUM MINING
Southwest Research & Information Center
P.O. Box 4524
Albuquerque, New Mexico 87106

WASTE
Southwest Research & Information Center
P.O. Box 4524
Albuquerque, New Mexico 87106

Regional Antinuclear Groups:

Abalone Alliance
452 Higuera
San Luis Obispo, California 93401
(803) 543-6614

Armadillo Coalition
Box 15556
Fort Worth, Texas 76116
(817) 926-1745

Catfish Alliance
362 Binkley Drive
Nashville, Tennessee 37415
(615) 832-0392

Citizens Against Nuclear Power
711 S. Dearborn
Chicago, Illinois 60605
(312) 764-5011

Citizens Against Nuclear Threats
106 Girard S.E.
Albuquerque, New Mexico 87106
(505) 268-9557

Clamshell Alliance
62 Congress Street
Portsmouth, New Hampshire 03801
(603) 436-5414

Crabshell Alliance
4337 Phinney Avenue
Seattle, Washington 98122
(206) 325-1983

Environmental Coalition on
    Nuclear Power
433 Orlando Avenue
State College, Pennsylvania 16801
(814) 237-3900

Northern Sun Alliance
1573 E. Franklin Avenue
Minneapolis, Minnesota 55404
(612) 874-1540

Palmetto Alliance
18 Bluff Road
Columbia, South Carolina 29201
(803) 254-8132

Rocky Flats Action Group
1432 Lafayette
Denver, Colorado 80218
(303) 832-1676

Trojan Decommissioning Alliance
215 S.E. 9th Avenue
Portland, Oregon 97214
(503) 231-0014